William Henry Maxwell

The Construction of Roads and Streets

With historical sketch of the development of the art of road-making

William Henry Maxwell

The Construction of Roads and Streets
With historical sketch of the development of the art of road-making

ISBN/EAN: 9783337012618

Printed in Europe, USA, Canada, Australia, Japan

Cover: Foto ©ninafisch / pixelio.de

More available books at **www.hansebooks.com**

By T. D. Scott.

JOHN METCALF.

JOHN LOUDON MACADAM.

THOS. TELFORD.

THE CONSTRUCTION
— OF —
ROADS AND STREETS,
— WITH —
HISTORICAL SKETCH
OF THE
DEVELOPMENT OF THE ART OF ROAD-MAKING.

BY

WILLIAM H. MAXWELL, C.E.,
Assistant Engineer and Surveyor,
LEYTON.

WITH NUMEROUS SPECIALLY-PREPARED ILLUSTRATIONS.

LONDON:
THE ST. BRIDE'S PRESS, LTD.,
24 Bride Lane and 13 New Street Hill, Fleet Street.
1899.

THE PIONEER

AUSTRALIAN HARDWOOD.

Sanitary, Durable, Non-Absorbent.

JARRAHDALE JARRAH.

Unexcelled for . . .

STREET PAVING, PLATFORM PLANKING, RAILWAY TRUCKS, HARBOUR WORK, RAILWAY SLEEPERS, STAIR TREADS, FENCE POSTS, PALINGS, &c., &c.

For quotations and particulars apply to the sole importers,

The Jarrahdale Jarrah Forests and Railways, Limited.

Agency Offices:—**1 Fenchurch Avenue, LONDON, E.C.**

PREFACE.

THESE notes have been prepared, at the request of the St. Bride's Press, in order to embrace the various branches of the subject of the "CONSTRUCTION OF ROADS AND STREETS" as required in the ENGINEERING SECTION of the Examination of the INCORPORATED ASSOCIATION OF MUNICIPAL AND COUNTY ENGINEERS, and it is hoped they may be found to adequately meet this purpose, as well as to supply such practical information upon the subject of modern Road Construction and Maintenance as will be of interest to Municipal Engineers and Surveyors having charge of these matters.

WILLIAM H. MAXWELL.

TOWN HALL, LEYTON.
1899.

FLAGSTONE, KERB, EDGING AND CHANNELS,

Of the most durable character, bright in colour, non-slippery, and composed almost entirely of silicates, for FOOTWAYS having constant wear in Cities and Towns.

SETTS, FOR CARRIAGEWAYS, CROSSINGS, &c.

SUPPLIED BY

THE LUOGH SOUTH QUARRY CO.,
LIMITED.

REGISTERED OFFICE—	QUARRIES—
Bury Road, Rawtenstall, Manchester.	**Liscannor, near Lahinch, Co. Clare, Ireland.**

The QUARRY is located on the MILLSTONE GRIT SERIES, on the same geological horizon as the celebrated FLAGSTONE QUARRIES OF HASLINGDEN AND ROCHDALE, which supply the Cities and Towns of MANCHESTER, LIVERPOOL, ST. HELENS, BOLTON, STOCKPORT, OLDHAM, ROCHDALE, BURY, WARRINGTON, BLACKBURN, BLACKPOOL.

Co. Clare STONE is largely exported to LONDON, CARDIFF, NEWPORT, SWANSEA, LLANELLY, BIRKENHEAD, Belfast, Dublin, Cork, Limerick, &c.

Managing Director—Mr. JOHN H. SPENCER, F.G.S.   Secretary—Mr. A. HOLT.

THE MUNICIPAL OFFICER.

The Official Journal of the Municipal Officers' Association.

MONTHLY, 6d. (free to members of the association).
YEARLY, 6s. (post free).

Editorial Offices: ST. BRIDE'S HOUSE, 24 BRIDE LANE, FLEET STREET, E.C.
Publishing Office: 13 NEW STREET HILL, FLEET STREET, E.C.

ROAD MATERIALS.

MOUNTSORREL GRANITE CO.,
LIMITED.

Quarries:

MOUNTSORREL, near Loughborough, on the Main Line of the Midland Railway.

STONEY STANTON, on the South Leicestershire Branch of the London and North-Western Railway.

Dressed Masonry, Millstones and Rollers, Kerb, Paving Setts, Randoms or Rough Setts, Macadam of various sizes, Concreters' Material, Granite Sand, &c., &c., delivered to all parts of the kingdom by rail or water.

FULL PARTICULARS AND PRICES ON APPLICATION.

TABLE OF CONTENTS.

	PAGE
HISTORICAL SKETCH OF ROAD-MAKING	1 to 23
SELECTION OF LINES AND LEVELS OF A NEW ROAD	23
THE BEARING OF GEOLOGY UPON ENGINEERING WORKS	29
GENERAL OBSERVATIONS AS TO ROUTE OF LINE OF ROADWAY, INCLINATION, &C.	31
RIVERS AFFECTING ROUTE	34
CROSSING EXTENSIVE PLAINS, MARSH GROUND, &C.	35
EARTH WORK, EMBANKMENTS, CUTTINGS, AND THEIR DRAINAGE	38
COMPUTATION OF QUANTITIES OF EARTH WORK	48
ROAD DRAINAGE	51
FENCING OF COUNTRY ROADS, &C.	64
TRACTION ON ROADS	66
WIDTH, CROSS SECTION AND GRADIENT OF ROADS	79
ROAD MATERIALS AND CONSTRUCTION	88
CONSTRUCTION OF ROADS	109
CONCRETE MACADAM	122
TAR MACADAM	125
COST OF CONSTRUCTION OF MACADAMISED ROADWAYS	128
ROAD-ROLLING	131
TOWN ROADS	140
STONE PAVEMENTS	142
GRANITE PITCHED ROADS IN LIVERPOOL	145
BRICK PAVEMENTS	153
WOOD PAVEMENTS.	161
WOODS EMPLOYED IN STREET PAVING	164
CREOSOTING TIMBER	170
CONSTRUCTION OF WOOD PAVEMENTS AND VARIOUS SYSTEMS OF LAYING	179
JOINTING OF WOOD PAVEMENTS	186
ASPHALTE PAVEMENTS	196

	PAGE
SPECIFICATION FOR COMPRESSED ASPHALTE ROADWAYS	202
FOOTWAYS ...	203
CURBING AND CHANNELLING	207
MATERIALS USED FOR PAVING FOOTWAYS ...	213
NATURAL STONE USED FOR PAVING	214
SPECIFICATION FOR YORK PAVING	216
ARTIFICIAL PAVING STONE	218
CONCRETE PAVING LAID "IN SITU"	222
BRICK PAVEMENTS	225
TAR PAVEMENTS	226
ASPHALTE FOR FOOTWAYS	228
CORK PAVEMENTS	231
MOVING PAVEMENTS	232
SELECTION OF PAVING MATERIALS	232
THE CONSTUCTION OF NEW STREETS UNDER THE BY-LAWS OF THE LOCAL GOVERNMENT BOARD ...	235
MAKING UP OF STREETS UNDER THE PRIVATE STREETS WORKS ACT, 1892	242
PRIVATE STREET WORKS ACT, 1892 ...	245

LIST OF ILLUSTRATIONS.

FIG.	DESCRIPTION.	PAGE
	John Metcalf	Frontispiece
	John Loudon Macadam	,,
	Thomas Telford	,,
1.	Cross Section of a French Road previous to 1764	4
2.	Cross Section of a French Road as adopted by M. Tresaguet (1764)	4
3.	English Road about 1809 ("Barrelled")	11
4.	Cross Section of Telford's Roads	16
5.	Diagram Illustrating Telford's Method of Surveying by Rectangular Lines	25
6.	Method of Embanking, with Line crossing Track of Flat Country Subject to Floods	35
7.	Method of Embanking through Soft Ground	37
8.	Embankments formed Concavely	40
9.	Embankments formed Convexly	40
10.	Embanking upon "Side-long" Ground	44
11.	Roadway upon "Side-long" Ground, showing "Catch-Water" Drain and Disposal of Surface Water	45
12.	Roadways upon "Side-long" Ground, with Retaining Walls	47
13.	Calculation of Irregular Figure by Simpson's Rule	50
14.	Cross Section of 30-ft. Country Main Road	53
15.	Stone Side Drain for Surface Drainage	54
16.	Plan of Road, showing Cross Drains from Centre of Road to Side Main Drain	55
17.	Cross or Mitre Drains	56
18.	Cross Section of "Inlet" to carry Water from Side Channel to Cross Drain	57

		PAGE
19.	Longitudinal Section of Cross Drain, showing "Inlet" from Side Channel ...	58
20.	Another Mode of Removal of Surface Drainage from Roads	59
21.	Removal of Surface Drainage from Roads (alternative arrangements)	59
22.	Cross Section of a Town Road, showing General Arrangements of Sewers and Connections for Removal of Surface Water	60
23.	Stoneware Circular Street Gully, showing Fixing	64
24.	Traction on Inclined Plane	71
25.	Cross Section of Suburban Macadamised Roadway	85
26.	Residential Road	115
27.	Business Road	116
28.	Cross Section of French Road	123
29.	Cross Section of a First-Class Macadam Road	128
30.	Suburban Macadam Road	129
31.	Road-rolling (Cross Section of Roadway) ...	135
34.	First-Class Granite Pitched Streets, Liverpool	147
35.	Second ,, ,, ,, ,, ,,	147
36.	Third ,, ,, ,, ,, ,,	149
37.	Fourth ,, ,, ,, ,, ,,	149
38.	Fifth ,, ,, ,, ,, ,,	152
39.	Cross Section of a Wood-paved Street ...	187
40.	Contour for Wood-paved Roads	192
41.	Curbing, Channelling, &c.	204
42.	,, ,, ,, ...	205
43.	,, ,, ,, ...	206
44.	Fire-Clay Brick Curb	207
45.	Bull-Nose ,, ,, ...	208
46.	Bull-Nose ,, ,, ...	208
47.	Splayed Fire-Clay Brick Curb	210
48.	Iron Curbs	211
49.	Cast-Iron Curbs	212

W. H. HARING'S
Drawing Instruments

Are acknowledeged by all who use them to be the BEST IN THE MARKET.

VERY GREAT CARE is taken in their manufacture, and **EVERY PIECE** is examined by a competent workman before leaving the Factory.

The following letter confirms the above:—

Moss Side, Exmouth, 26th Nov, 1898.

"Herewith please find case of instruments purchased from you some twenty odd years ago, and kindly clean and repair same and complete those missing, at your earliest convenience.

"Allow me to say I never wish to work with better instruments than they have been.

"Yours truly,
"W. H. Beswick, A.M.I.C.E., &c."

W. H. HARING,
47 Finsbury Pavement, London, E.C.

Contractor to H.M. War Department, Admiralty, Council of India, &c.

Regd. Telegraphic Address: "CLINOGRAPH, LONDON."

ESTABLISHED 1851.

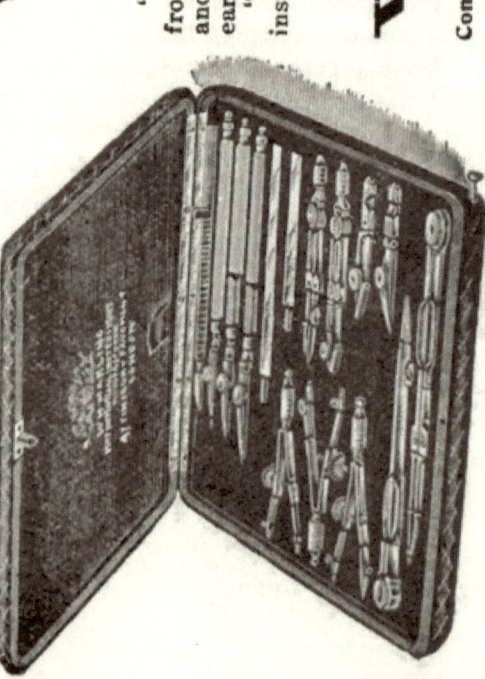

THE CONSTRUCTION OF ROADS AND STREETS.

HISTORICAL SKETCH.

Before entering upon the discussion of the design, construction and maintenance of modern roads and streets it will be advantageous to review concisely the history of the evolution of this very important branch of civil engineering practice, illustrating very forcibly, as it does, and probably more universally so than any other branch of engineering work, Mr. Tredgold's definition* of the profession of the civil engineer—*viz.*, "The art of directing the great sources of power in nature for *the use and convenience of man.*"

Good *roads* are among the most influential agencies of society, and *road-makers* have proved the most effective pioneers of civilization. By opening up such lines of communication the city and the town are brought into connection with the village and the farm, markets are found for field produce, and outlets provided for manufacturers. Thus the natural resources of the country are developed, travelling, intercourse, industry and commerce are rapidly set on foot, and, in fact, "the road is so necessary an instrument of social well-being that in every new colony it is *one of the first things thought of.* . . The new country, as well as the old, can only be effectually 'opened up,' as the common phrase is, by *roads*, and until these are made it is virtually closed."†

The invention of *paved roads* is given to the Carthaginians. Their example was afterwards adopted by the Romans,‡ who

---

* Adopted by the Institution of Civil Engineers in their charter.
† Smiles' "Lives of the Engineers," vol. i.
‡ See French Encyclopædia.

B

were the first to bring the art of *road-making* into England. The roads of ancient Rome are the earliest about which we have any accurate knowledge, one of the oldest and most important of these being the great paved military highway called the "Appian Way" (*Via Appia*), which was constructed partly by *Appius Claudius*, about 312 B.C. This road extended from Rome to Capua, a distance of 120 miles, and was long afterwards continued in a south-easterly direction across the Appennines to Brundusium (now Brindisi), in the south of Italy.

"*Roman roads* are remarkable for preserving a straight course from point to point, regardless of obstacles which might have been easily avoided. They appear to have been often laid out in a line with some prominent landmark, and their general straightness is perhaps due to convenience in setting them out. In solidity of construction they have never been excelled, and many of them still remain, often forming the foundation of a more modern road, and in some instances constituting the road surface now used."*

The mode of construction was as follows: "Two parallel trenches were first cut to mark the breadth of the road; loose earth was removed until a solid foundation was reached; and it was replaced by proper material consolidated by ramming, or other means were taken to form a solid foundation for the body of the road. This appears, as a rule, to have been composed of four layers, generally of local materials, though sometimes they were brought from considerable distances. The lowest layer consisted of two or three courses of flat stones, or, when these were not obtainable, of other stones, generally laid in mortar; the second layer was composed of rubble masonry of smaller stones, or a coarse concrete; the third of a finer concrete, on which was laid a pavement of polygonal blocks of hard stone *jointed with the greatest nicety*. The four layers are found to be often 3 ft. or more in thickness, but the two lowest were dis-

---

* "*Encyclopædia Britannica*," article on "Roads and Streets" by Thomas Codrington, C.E.

pensed with on rock. The paved part of a great road appears to have been about 16 ft.* wide, and on either side, and separated from it by raised stone causeways† were unpaved sideways, each of half the width of the paved road. Where, as on many roads, the surface was not paved, it was made of hard concrete, or pebbles or flints set in mortar. Sometimes clay and marl were used instead of mortar, and it would seem that where inferior materials were used the road was made higher above the ground and rounder in cross-section. Streets were paved with large polygonal blocks laid as above described, and footways with rectangular slabs. Specimens are still to be seen in Rome and Pompeii."‡ In flat districts the middle part of the roads was raised into a terrace¶ above the adjacent country.

"The public and the senate held the roads in such estimation, and took so great an interest in them, that under Julius Cæsar the principal cities of Italy all communicated with Rome by *paved roads*. Their roads from that period began to be extended into the provinces."§

"One of the grand causes of the civilisation introduced by that ruling people into the conquered states was the highways, which form, indeed, *the first germ of national industry*, and without which neither commerce nor society can make any considerable progress. Conscious of this truth, the Romans seem to have paid particular attention to the *construction of roads* in the distant provinces; and those of *England*, which may still be traced in various ramifications, present a lasting monument of the justice of their conception, the extent of their views, and the utility of their power. A grand trunk, as it may be called, passed from the south to the north, and another to the west, with branches in almost every direction that general convenience and expedition could require. What is called '*Watling-street*' led from Richborough,

---

* Roman feet.
† Two feet wide.
‡ "*Encyclopædia Britannica*."
¶ Tredgold on "Railways."
§ French Encyclopædia.

B²

in Kent, the ancient Rutupiæ, north-east, through London to Chester. The '*Ermine-street*' passed from London to Lincoln, thence to Carlisle, and into Scotland. The *Foss Way* is supposed to have led from Bath and the western regions, north-east, till it joined the Ermine-street. The last celebrated road was '*Ikeneld*,' or '*Ikneld*,' supposed to have extended from near Norwich, southward, into Dorsetshire."[*]

"There are no traces of Roman influence in the later roads in England, but in France the Roman method appears to have

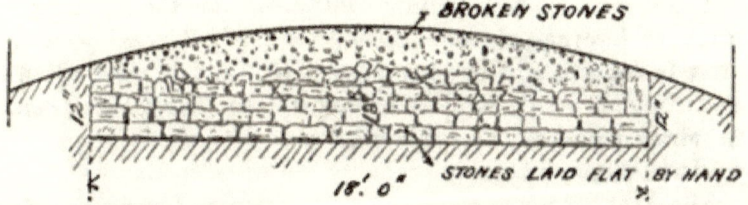

FIG. 1.—CROSS SECTION OF A FRENCH ROAD PREVIOUS TO 1764.

been followed to some extent when new roads were constructed about the beginning of the eighteenth century. A foundation of stones on the flat was laid (*See* Fig. 1), and over that two layers of considerable thickness, of larger and

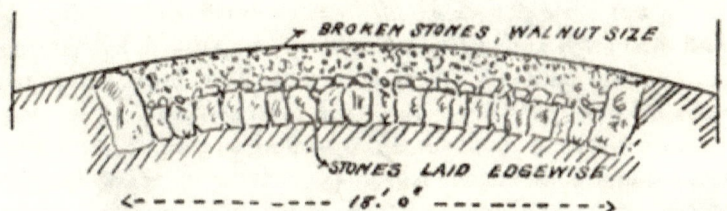

FIG. 2.—CROSS SECTION OF A FRENCH ROAD AS ADOPTED BY M. TRESAGUET (1764).

Rise of cross section about 6 in. at the crown or 1 in 36.

smaller stones, bordered by large stones on edge, which appeared on the surface of the road. In 1764[†] Trésaguet set the foundation-stones on edge (Fig. 2) and reduced the thick-

---

[*] "Pinkerton's Geography," vol. i., p. 20.

[†] French roads were, down to the year 1764, maintained by statute labour, and the repairs effected only twice annually, in the spring and autumn, thus necessitating a thickness of 18 in. at the middle and 12 in. at the sides. After this date the average thickness was reduced to about 10 in. throughout the cross-section.

ness of the upper layers, and his method was generally followed until the influence of Macadam began to be felt."*

French road-makers a century ago were very much ahead of their English contemporaries. This will be evident upon comparing the cross-section of the English road of 1809 with that of the French road of about 1775.

Macadam's principles of road construction were, it is stated, officially adopted in France about 1830, and became general not many years later.

The Roman roads in Britain continued to be the main highways of internal communication for a considerable period after the occupation of that people; but, having been subsequently neglected and allowed to fall into decay, they were in most instances overrun with forest and waste, until the roads of England became probably the worst in Europe.

Thus the means of communication between towns in early times were both difficult and dangerous; the roads or ways consisted of mere horse-tracks or footways across the country, and were traversed chiefly by pack-horses, which then formed the principal means for the transport of inland commerce from town to town.

"In very early periods of English civilisation, while the population was thin and scattered and men lived by hunting and pastoral pursuits, the track across the down, the heath and the moor sufficiently answered their purpose. Yet, even in those districts unencumbered with wood, where the first settlements were made—as on the downs of Wiltshire, the moors of Devonshire and the wolds of Yorkshire—stone tracks were laid down by the tribes between one village and another. In some districts they are called trackways or ridgeways, being narrow causeways usually following the natural ridge of the country, and probably serving in early times as local boundaries—on Dartmoor they are constructed of stone blocks, irregularly laid down on the surface of the ground, forming a rude causeway of about 5 ft. or 6 ft. wide."†

---

* "*Encyclopædia Britannica.*"
† Smiles' "Lives of the Engineers," vol. i.

As to the origin of the present *lines* of roadways, the following is suggested in a pamphlet, entitled "The Landed Property of England": "Most of the old roads of the kingdom (the **remains** of the Roman ways excepted) owe their present lines to particular circumstances—many of them were, no doubt, originally footpaths; some of them, perhaps, the tracks of the aboriginal inhabitants; and these footpaths became, as the condition of society advanced, the most convenient horse-paths. According as the lands of the kingdom were appropriated the tortuous lines of road became fixed and unalterable, there being no other legal lines left for carriage roads, and hence the origin of the crookedness and steepness of existing roads."

"In some of the older-settled districts of England the old roads are still to be traced in the *hollow ways* or lanes which are met with, in some places 8 ft. and 10 ft. deep. Horse tracks in summer and rivulets in winter, the earth became gradually worn into these **deep furrows**, many of which, in Wilts, Somerset and **Devon**, represent the tracks of roads as old as, if not older **than**, the Conquest. When the ridgeways of the earliest **settlers on** Dartmoor, above alluded to, **were** abandoned, the tracks were formed through the valleys, **but** the new roads were no better than the old ones. They were narrow and deep, fitted only for a horse passing along laden with its crooks, as so capitally described in the ballad of '*The Devonshire Lane.*'"*

"In the neighbourhood of London **there was a** *hollow way* which now gives its name to a populous metropolitan parish. Hagbush-lane was another of such roads; before the formation of the Great North Road it was one of the principal bridle-paths leading from London to the northern parts of England; but it was so narrow as barely to afford passage for more than a single horseman, and so deep that the rider's head was beneath the level of the ground on either side."†

---

* This ballad amusingly describes *marriage* as being "just like a Devonshire lane," into which, the traveller having once entered "there is no turning round," &c.

† Smiles' "Lives of the Engineers."

Attempts were, however, made from time to time for the preservation of the highways, the state of which may be inferred from the following: "One of the first laws on the subject was passed in 1285, directing that all bushes and trees along the roads leading from one market to another should be cut down for 200 ft. on either side, to prevent robbers lurking therein; but nothing was proposed for amending the condition of the ways themselves. In 1346 Edward III. authorised the first toll to be levied for the repair of the roads leading from St. Giles-in-the-Fields to the village of Charing (now Charing Cross), and from the same quarter to near Temple Bar (down Drury-lane), as well as the highway then called Perpoole (now Gray's Inn-lane). The footway at the entrance of Temple Bar was interrupted by thickets and bushes, and in wet weather was almost impassable. The roads further west were so bad that when the sovereign went to Parliament faggots were thrown into the ruts in King-street, Westminster, to enable the royal cavalcade to pass along."

In the first Act (passed in 1532) for paving and improving the City of London the streets were described as "very foul, and full of pits and sloughs, so as to be mighty perillous and noyous, as well for all the king's subjects on horseback as on foot with carriages."

"The first *attempt* to put the roads of England into order occurred when the *turnpike system* was introduced. The ancient method employed to mend roads, until after the restoration of King Charles II., was by a £1 rate on the land-holders in the respective counties, and by the supply of carts and horses by parishes for a limited number of days. But when, after the last-named period, commerce so generally increased, and in consequence thereof wheel-carriages and pack-horses were extremely multiplied, *the first turnpike road* was established by law (the 16 Charles II., cap. 1., *anno* 1653) for taking toll of all but foot passengers on the northern road through Hertfordshire, Cambridgeshire and Huntingdon-shire, which road was then become very bad by means of the great loads of barley and malt, &c., brought weekly to Ware in waggons and carts, and from thence conveyed by water to London."*

Hume, in his "History of England," speaking of the "internal state of England" during the reign of James II. (1685 to 1688), says: "The means of communication throughout the kingdom were wretched in the extreme. Canals did not exist; the roads were execrable and infested with highwaymen. Four horses, sometimes six, were required to drag the coaches through the mud; and the traveller who missed the scarce discernible track over the heaths, which were then frequent and extensive, might wander lost and benighted. Some improvement was effected by the introduction of posts in the reign of Charles I., which were brought to more perfection after the Restoration. . . . . But no very considerable improvement in the roads took place till the reign of George II.," when the Rebellion in Scotland (1745) gave considerable impulse to the construction of roads for military as well as for civil purposes. After its suppression the Government directed attention to the best means of securing permanent subordination of the Highlands, and with this object the construction of good highways was considered indispensable.

"Even till towards the end of last century the roads in many parts of England were execrable. The best coaches on a long journey cleared no more than 4 or 5 miles an hour. After the Peace (of Paris, 1815) the roads were very much improved by the use of broken stones and granite introduced by Macadam, and the pace was in many instances accelerated to 10 miles an hour."

"The almost incredibly bad state of the roads in England towards the latter part of the seventeenth century also appears from the accounts cited by Macaulay (Hist., c. iii.). It was due chiefly to the state of the law, which compelled each parish to maintain its own roads by statute labour; but the establishment of turnpike trusts and the maintenance of roads by tolls do not appear to have effected any great improvement."†

---

* "A Treatise on Roads," by Sir Henry Parnell.
† "*Encyclopædia Britannica.*"

Little or no thought was given to the scientific construction of roads, and the persons engaged to make and keep them in repair were altogether incompetent and ignorant of the most elementary principles involved in their proper construction. "Road-making as a profession was as yet unknown. . . . Men of eminence as engineers—and there were very few such at the time—considered road-making beneath their consideration; and it was even thought singular that, in 1768, the distinguished Smeaton should have condescended to make a road across the valley of the Trent between Markham and Newark."* The adoption of a good system of *road-making*, like most great reforms, was a very gradual process, and grew only with the education of the public mind as to the importance of having better and more rapid means of communication.

The introduction of "*stage coaches*" about the middle of the seventeeth century formed a new era in the history of travelling by road. They were at first no better than a sort of waggon, and the pace did not exceed 4 miles an hour. The jolting to which the passengers were treated in the earlier vehicles without springs and travelling over such bad roads must have been very considerable.† Probably the first coaches for public accommodation were run between London and Dover. Their introduction was regarded with much prejudice, and had much opposition to encounter.

An idea of the state of some of the roads in the north of England may be gained from Arthur Young's "Six Months' Tour," published in 1770, from which it appears that up to that time the roads had undergone little improvement. He says: "*To Wigan Turnpike.* I know not, in the whole range of language, terms sufficiently expressive to describe this infernal road. Let me most seriously caution all travellers, who may accidentally propose to travel this terrible country, to avoid it as they would the devil, for a thousand to one they break their necks or their limbs by overthrows or breakings

---

* Smiles' "Lives of the Engineers."
† It used to be said of the *drivers* that they were " seldom sober, never civil, and always late!"

down. They will here meet with ruts, which I actually measured 4 ft. deep, and floating with mud only from a wet summer; what, therefore, must it be after a winter? The only mending it receives is tumbling some loose stones, which serve no other purpose than jolting a carriage in the most intolerable manner. These are not merely opinions, but facts; for I actually passed three carts broken down in those 18 miles of execrable memory."

In "1736 we find Lord Hervey, writing from *Kensington*, complaining that 'the road between this place and *London* is grown so infamously bad that we live here in the same solitude as we would do if cast on a rock in the middle of the ocean; and all the Londoners tell us that there is between them and us an impassable gulf of mud.' The mud was no respecter of persons either, for we are informed that the carriage of Queen Caroline could not, in bad weather, be dragged from St. James's Palace to Kensington in less than two hours, and occasionally the royal coach stuck fast in a rut, or was even overthrown into the mud. The *streets of London* themselves were no better at that time, the kennel being still permitted to flow in the middle of the street, which was paved with *round stones*, flagstones for the pedestrians being as yet unknown."*

The extension of the *turnpike system* encountered violent opposition from the people, especially in Yorkshire, Somersetshire and Gloucestershire. Armed bodies of men assembled to destroy the turnpikes, and they burnt down the toll houses and blew up the posts with gunpowder. The country people in some places would not even use the improved roads after they were constructed†; but the opposition did not retard the progress of the turnpike and highway legislation, as may be seen from the fact that from 1760 to 1774 no less than 452 Acts were passed for making and repairing highways.

---

* Smiles' "Lives of the Engineers."
† The Blanford waggoner is reported to have said: "Roads have but one object—for waggon driving. He required but 4-ft. width in a lane, and all the rest might go to the devil." He continued: "The gentry ought to stay at home and be d——d, and not run gossiping up and down the country." Robert's "Social History of the Southern Counties."

From the "Third Report from Parliamentary Committee on Turnpikes and Highways, 1809," it appears that a highly convex cross-section (Fig. 3) was at that time generally adopted, with the view of keeping the road dry; but the convexity was usually so great as to render the sides dangerous for the passage of traffic. Vehicles, therefore, kept entirely to the centre, and deep ruts, retaining large quantities of water and mud, very soon resulted.

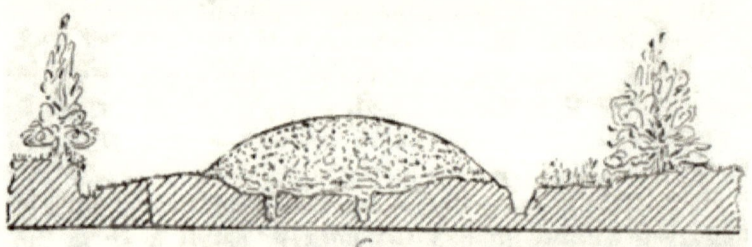

FIG. 3.—ENGLISH ROAD ABOUT 1809,
Showing its state before and after repair. Well "barrelled!"

The first real improvement in the business of *road-making*, and the first application of scientific principles to their construction and repair, is associated with the names of Metcalf, Macadam and, more especially, Telford.

Metcalf's Roads.—John Metcalf, born at Knaresborough in 1717, was the son of poor working parents, and, though totally blind from the age of six years, he was possessed throughout his career of extraordinary activity, spirit of enterprise, and a daring boldness. After having been successively a fiddler, soldier, chapman, fish dealer, horse dealer and waggoner, he entered upon the main business of his remarkable life—viz., that of a *road and bridge constructor*. He constructed numerous important turnpike roads throughout Yorkshire, Lancashire, Cheshire and Derby, amounting to a total mileage of about 180 miles, involving the building of many bridges, retaining walls, culverts, &c. He also successfully crossed several difficult bogs, by putting down large quantities of furze and ling bound together in little round bundles, and laid in longitudinal and transverse layers alternately,* and then spreading coatings of stone and gravel over same.

Macadam's roads are so named after John Loudon Macadam, who about the year 1816 undertook the management of the roads in the Bristol district. The distinguishing characteristic of his system was, to use Mr. Macadam's own description, "to put broken stone upon a road, which shall unite by its own angles so as to form a solid, hard surface. It follows that when that material is laid upon the road it must remain in the situation in which it is placed without being moved again; and what I find fault with in putting quantities of gravel on the road is that, before it becomes useful, it must move its situation and be in constant motion."†

Macadam's practice was to lay the "*road metal*" directly upon the natural ground, the only preparation being the levelling of inequalities and the digging of side drains. The maximum thickness of metalling adopted was about 10 in. in depth, and consisted of a layer of flints, or some other hard material, broken to a uniform size, of approximately cubical shape, about 1½ in. to 2 in. in diameter, and weighing not more than 6 oz.‡ The broken stone was spread evenly over the road surface and consolidated by the traffic, the whole surface being regularly raked over and levelled during consolidation. Additional materials were added, as found necessary to maintain an even surface, in thin uniform layers, either after rain or after a thorough moistening with water, thus insuring a complete incorporation of the old and new metal.

Macadam strongly condemned the use of any "*binding material*" for filling the interstices in the metal (a practice now universally adopted), which he left to work in and unite by its own angles by means of the traffic.

---

* A similar plan was adopted by George Stephenson when constructing a railway across Chat Moss.

† "Report of the Select Committee on the Highways of the Kingdom, 1819."

‡ "Macadam directed each road inspector to carry a small balance, so as to be able to test the weight of a few stones from each heap." (*Rankine.*)

He adopted only sufficient convexity of cross section as was necessary to keep the road dry by drainage of rain water to the side channels.

Macadam appears to have been connected more with the repairs of old roads than with the making of new ones. He first used his system in Ayrshire, where his attention was directed to the subject while acting as one of the trustees of a road in this county.

In 1825, having proved before a committee of the House of Commons an expenditure of several thousand pounds from his own resources in carrying out his improvements, the amount was reimbursed to him with an honorary tribute of £2,000.*

It is said, however, that "he did not *invent* the method in question of breaking stone, because it had long been the practice of Sweden, Switzerland and other countries, and was long know to every observing traveller."† Also the system was advocated by a Mr. Edgeworth in "An Essay on the Construction of Roads and Carriages," a *second* edition of which was published in 1817.

The term "macadamised roads," strictly speaking, does not apply to roads having a pitched foundation, but only to those constructed wholly of broken stones. The expression now, however, is generally understood to include all roads composed of, or repaired with, broken stone.

Telford's roads.—Telford first gave attention to the construction of roads in 1803-4; his practice lay more with the construction of new roads than with the repair of old ones. He constructed the high road from London to Holyhead and Liverpool, many miles of roads in the Scottish Highlands and elsewhere in the North. His special object was to ensure the entire separation of the broken stone or "*road metal*" from the subsoil, and for this purpose he first laid down "*a pitched foundation*" or "*bottoming*," "consisting of pieces of durable, but

---

* Smiles' "Lives of the Engineers."
† *The Westminster Review*, vol. iv., p. 354.

not necessarily hard, stone measuring from 4 in. to 7 in. in each dimension. The largest of those pieces are set by hand, with their largest sides resting on the '*formation*,' and between these the smaller pieces are packed, so as to form a compact layer about 7 in. deep in the centre of the road and 4 in. deep at the sides, part of the convexity being made in this manner."* Upon the foundation was laid a thickness of road metal of broken stone similar to Macadam's, which decreased in size towards the finished surface.

Telford looked to thorough drainage, the use of suitable materials, and, instead of the exaggerated roundness above referred to, he adopted for the form of cross-section of the convexity of the carriageway a very flat ellipse.

Attention was also paid to the setting out of new roads, to the rates of inclination,[†] and to the acclivities to be attained —avoiding all needless ascents—so that traction may be effected with the minimum expenditure of power. As an instance of Telford's skill in road engineering, the case of the old road, 24 miles in length, across the island of Anglesea may be cited, which was so undulated that a horse ascended and descended 1,283 ft. (vertical height) more than was found to be necessary by Telford. Also the line of his new road was 2 miles 592 yards shorter than the old one. Another example of the serious injury which the public sustain through the lack of engineering skill in the laying-out of roads may be found in the case of the road between London and Barnet, " on which the total number of perpendicular feet that a horse must now ascend is upwards of 1,300, although Barnet is only 500 ft. higher than London; and in going from Barnet to London a horse must ascend 800 ft., although London is 500 ft. lower than Barnet."‡

Telford's system of road-making was afterwards followed by his assistant, Mr. (afterwards Sir John) Macneil, and there is little or no essential difference between that method and the practice followed at the present day.

---

* " Manual of Civil **Engineering**," by Prof. W. J. Macquorn Rankine (C. Griffin & Co.).
† Some of which he reduced from about 1 in 6 to a minimum of 1 in 20.
‡ " A Treatise on Roads," by Sir Henry Parnell.

The following *specification of a 30-ft. roadway* as designed by Telford* will be of interest†: "Upon the level bed prepared for the road materials a bottom course or layer of stones is to be set by hand in form of a close, firm pavement; the stones set in the middle of the road are to be 7 in. in depth; at 9 ft. from the centre 5 in.; at 12 ft. from the centre 4 in.; and at 15 ft. 3 in. They are to be set on their broadest edges lengthwise across the road, and the breadth of the upper edge is not to exceed 4 in. in any case. All the irregularities of the upper part of the said pavement are to be broken off by the hammer, and all the interstices to be filled with stone chips firmly wedged or packed by hand with a light hammer, so that when the whole pavement is finished there shall be a convexity of 4 in. in the breadth of 15 ft. from the centre.

"The middle 18 ft. of pavement is to be coated with hard stones to the depth of 6 in. Four of these 6 in. to be first put on and worked in by carriages and horses, care being taken to rake in the ruts until the surface becomes firm and consolidated, after which the remaining 2 in. are to be put on.

"The whole of this stone is to be broken into pieces as nearly cubical as possible, so that the largest piece in its longest dimensions may pass through a ring of $2\frac{1}{2}$ in. inside diameter.

"The paved spaces on each side of the 18 middle feet are to be coated with broken stones or well-cleaned stony gravel up to the footpath or other boundary of the road, so as to make the whole convexity of the road 6 in. from the centre to the sides of it, and the whole of the materials are to be covered with a *binding* of $1\frac{1}{2}$ in. of good gravel free from clay or earth" (*See* Fig. 4).

The Select Committee of the House of Commons, in reporting as to the manner in which the works in connection with the Holyhead-road had been carried out, stated as follows: "The professional execution of the new works upon this road

---

* "A Treatise on Roads," by Sir Henry Parnell.
† For a portion of the Holyhead road.

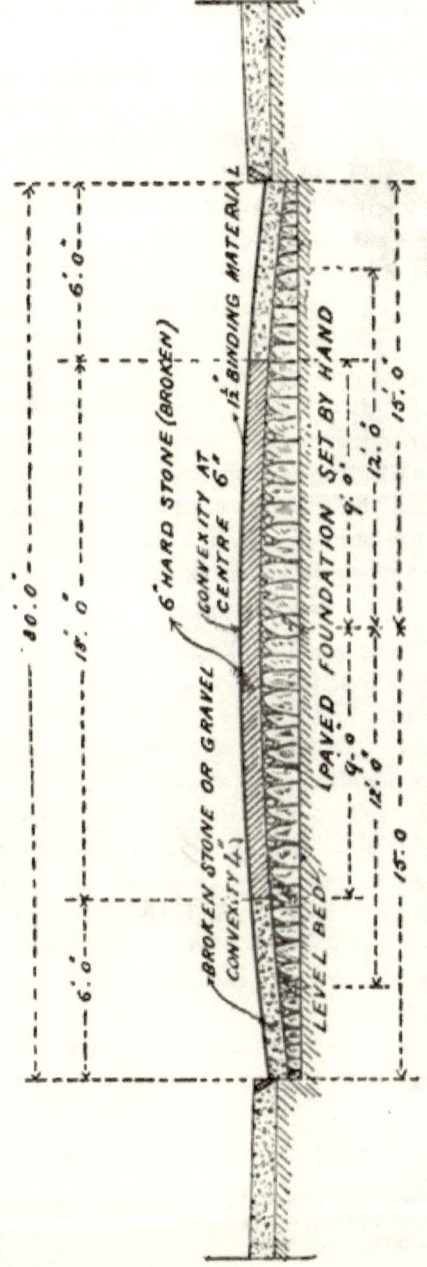

FIG. 4.—CROSS SECTION OF TELFORD'S ROADS, AS SPECIFIED.

greatly surpasses anything of the same kind in these countries. The science which has been displayed in giving the general line of the road a proper inclination through a country whose whole surface consists of a succession of rocks, bogs, ravines, rivers and precipices, reflects the greatest credit upon the engineer who has planned them; but perhaps a still greater degree of professional skill has been shown in the construction, or rather the building, of the road itself. The great attention which Mr. Telford has bestowed to give to the surface of the road one uniform and moderately convex shape, free from the smallest inequality throughout its whole breadth; the numerous land drains, and, where necessary, shores and tunnels of substantial masonry, with which all the water arising from springs or falling in rain is instantly carried off; the great care with which a sufficient foundation is established for the road; and the quality, solidity and disposition of the materials that are put upon it, are matters quite new in the system of road-making in these countries."*

ROAD-ROLLING.

The application of steam rolling, as we now know it, for the proper consolidation of macadam roads has, like most other improvements enjoyed at the present day, been the subject of gradual evolution. An interesting history of this subject, including *horse* road-rolling and *steam* road-rolling, is given in a report,† addressed to the Metropolitan Board of Works, 1870, by Mr. Frederick A. Paget, c.e., and according to which the first recorded allusion to road-rolling (*horse*-power, of course) seems to have been made in the letters patent granted in 1619 to a certain John Shotbolte, who speaks of using certain "strong and massy engines in the making and repairing of highways and roads."

"The first proposal of a road-roller, on a sound and scientific basis, was made by a French royal engineer of roads; and

---

* "Report from the Select Committee on the Road from London to Holyhead in the year 1819."
† "Report on the Economy of Road Maintenance and Horse Draft through Steam Road-rolling, with special reference to the Metropolis," by Frederick A. Paget, c.e., 1870.

in 1787 M. de Cessart, then *Inspecteur Général des Ponts et Chaussèes*, recommended a cast-iron roller for rolling down newly-metalled roads."

Patents for horse road-rollers were taken out in England by Philip Hutchinson Clay in 1817, and by John Biddle in 1825; and wooden rollers, filled with stones, sand or clay, are said to have been in use in France about 1840.

Although the rolling of macadam roads seems to have been applied in actual practice in 1830 or thereabouts, it appears to have been regarded somewhat as an unnecessary luxury in England, and was therefore not properly appreciated until about 1843, when, according to the report above mentioned, "the first recommendation in the English language of horse road-rolling as a measure of economy was published by Sir John F. Burgoyne, R.E.,[*] when chairman of the Board of Works of Ireland. He was one of the first engineers who used it in their own work, and certainly the first in England to scientifically recommend it, not as a refinement, but as a necessity."

"The first patent for a *steam road-roller* was taken out in France, at the beginning of 1859, by M. Louis Lemoine, of Bordeaux," and in 1863 Mr. W. Clark, chief engineer of the municipality of Calcutta, quite independently of the French inventors, "conceived the idea of a self-propelling steam road-rolling engine, with its weight uniformly distributed over the whole width of the rollers," and, in conjunction with Mr. W. F. Baths, of Birmingham, patented a design in 1863, which was that of the *first steam road-roller* ever tried or patented in Great Britain; and although "the first experiments were made in France, yet his steam roller was the first thoroughly successful implement of the kind."

Among other makers and patentees who appeared in the field before very long was the well-known firm of Messrs. Aveling & Porter, who "determined to adopt their form of traction engine to this purpose. The result is a combination

---

[*] Paper (written in 1843) "On Rolling New-made Roads," by General Sir John F. Burgoyne, Bart.

of their simple and efficient form of traction engine with the arrangement of rollers and turn-table patented in 1863 by Messrs. Clark & Baths." In 1867 a 30-ton steam roller was supplied to the borough authorities of Liverpool, "and was in use for some time, when it had to be abandoned, principally on account of the great width of the rollers, which would not conform to the contour of the carriageway,"* but its great weight was also, no doubt, found to be very undesirable.

The steam roller has now become to be almost universally used, and generally weighs from 10 to 15 tons, this weight having been found from experience to be the most suitable for the purpose.

STONE PAVEMENTS.

Large pebbles or rounded boulders from 6 in. to 9 in. in depth, and bedded in gravel or sand, or frequently upon the natural, formed the surface of early pitched roadways in the metropolis and other large towns. Such roads naturally presented very uneven surfaces, and the joints being very wide, large quantities of slush and filth were retained and ruts soon appeared under the influence of traffic.

As an advance upon the boulder pavement, *roughly-squared blocks of stone*, measuring from 6 in. to 8 in. across the surface were used; but these not being properly dressed made wide and irregular joints, and stones of varying breadths were placed in the same course, thus admitting water, which softened the subsoil and caused an irregular settlement of the stones under the influence of traffic.

In 1824 Mr. Telford was called upon to report† on the street pavements, &c., of the parish of St. George, Hanover-square, and he then pointed out the deficiencies of the existing system in the following terms: "The notorious imperfection of the carriageway pavement having been the cause of this report, it is needless to state that the surface is generally uneven, and not unfrequently sunk into holes, so as

---

* "Municipal and Sanitary Engineers' Handbook," by H. Percy Boulnois, c.e. (Messrs. E. & F. N. Spon).

† "This report is printed in Sir Henry Parnell's " Treatise on Roads" (second edition).

to render it not only incommodious, but dangerous to horses and wheel carriages.

"The causes of this imperfection are various, and of an extensive and serious nature.

"The stones, though generally of a tolerably good nature, are so irregular in their shape that even their surfaces do not fit; they almost universally leave wide joints, and instead of these joints being dressed square down from the surface, that is at right angles with the face, they more frequently come only in contact near the upper edges, and by tapering downward in a wedge-like form have their lower ends very narrow and irregular, leaving scarcely any flat base to bear weight.

"This form also unavoidably leaves a great portion of space between the stones, which the workmen fill with loose mould or other soft matter of which the bed or subsoil is composed.

"Another great defect is caused by inattention to selecting and arranging the size of stones; they are but too commonly so mixed that large and small surfaces are placed alongside of each other, and, acting unequally in support of pressure, create a continual jolting in wheel carriages, which, adding percussion to weight, is a powerful and destructive agent."

Also, the bed on which the stones are placed "has hitherto but too generally been formed of very loose matter, easily convertible into mud; and this matter, instead of being compressed by artificial means, has unavoidably been loosened by a sharp-pointed instrument, to suit the irregular depth and narrow bottoms, and to fill the chasms between the joints of the paving stones. From the width and irregularity of the joints, water easily sinks into and converts the before-mentioned soft matter into mud, which by the continual and violent action of carriage wheels is worked upon the surface and leaves the stones unsupported."

As to "the best mode of constructing pavements" in order to remedy these defects, he recommended the entire separation of the subsoil from the paving stones by the introduction of a good foundation or bed consisting of 6 in. of clean river ballast consolidated by the traffic before laying the pitchers, or, better, a bottoming 12 in. in depth of broken stones put on in layers 4 in. thick and rendered solid by the passage of traffic. Upon this bed the granite pitchers, worked flat on the face, and true and square on the sides in order to make close joints and to ensure the bottom and top superficies of the stones being equal, were then laid. The stones used were large, and varied in breadth from 4½ in. to 7½ in., in depth from 7 in. to 10 in., and length from 7 in. to 13 in.; also, they were selected, in order to place stones of equal breadth in the same course and so obtain a better joint.

Later experience has not been in favour of the use of such wide stones as those employed by Telford, as, although having a larger base, they do not appear to better support the weight and jolting of heavy traffic, but have a tendency to rock, to wear round at the angles, to become slippery, and do not afford a good foothold for horses.

The introduction of modern sett pavements is due to Mr. Walker, who in 1840 paved Blackfriars bridge with granite *setts** 3 in. broad and 9 in. deep, laid upon a bed of concrete 12 in. in thickness, and well jointed with mortar. Since this time the use of 3-in. setts has been generally followed, and a concrete foundation is now also usually adopted.

WOOD PAVEMENTS.

"Pavements formed of blocks of wood appear to have been first employed in Russia, where, according to the testimony of Baron de Bode,† it has been, though rudely fashioned, used for some hundreds of years."‡

Pavements of wood were laid down in New York on different systems by way of experiment during the year 1835-6, and the first wood pavement in London was laid in front of the Old Bailey in 1839, and was on the system of Mr. David Stead, who was the first to introduce these pave-

---

* Pitchers are known, according to their dimensions, either as "*setts*," "*cubes*" or "*blocks*." A stone measuring
6¼ in. by 3¼ in. by 5 in. to 7 in. long is termed a "*sett*."
4 in. by 4 in. by 4 in. long is termed a "*cube*."
4 in. by 4 in. by 6 in. deep is termed a "*block*."

† "Wood Pavement," by A. B. Blackie, 1842.

‡ "The Construction of Roads and Streets," by Henry Law and D. K. Clark.

ments into England. The Old Bailey pavement was laid without a proper foundation, and was not a success.

Stead's wood pavement, patented in 1838, consisted of hexagonal fir blocks, measuring from 6 in. to 8 in. across and 4 in. to 6 in. in depth, bedded on 3 in. of gravel (or sometimes on concrete), and laid upon a foundation previously prepared by levelling and ramming. The blocks were either chamfered on the edges or grooved across the face, in order to afford a better foothold. Stead also used *round blocks*, placed with the grain vertical, the interstices being filled with gravel or sand.

Other systems of wood paving were patented and tried in London about this time, but were not satisfactory.

Carey's wood pavement was one of the best of its time, was patented in 1839, and consisted of blocks measuring from $6\frac{1}{2}$ in. to $7\frac{1}{2}$ in. wide, 13 in. to 15 in. long, and about 8 in. or 9 in. in depth. The blocks were cut with projecting and re-entering angles, so as to interlock on all sides, the object being to prevent unequal settlement by distributing the pressure over adjoining blocks. This pavement was first laid in the Poultry in 1841, and in the same year in Mincing-lane, and in Gracechurch-street in the following year. The blocks were laid upon a layer of Thames ballast or sand on the old surface of the street, and were grouted with lime and sand.

As regards the width of wood blocks, similar experience has been gained as with stone pitching—*viz.*, that smaller blocks, of about 4 in. in width and of 6 in. in depth, answered better. Also lateral support by interlocking angles was found to be unnecessary.

The "*improved wood pavement*" was first used in London in the year 1871.

A vast amount of experience in the laying of wood pavements has been obtained in the streets of the metropolis and other large towns, and its construction has now reached a high degree of perfection.

ASPHALT PAVEMENTS.

According to the *Century Magazine*[*] asphalt first attracted

---

[*] Vol. xlvi., p. 903.

attention in Europe as a material suitable for paving in 1849. "It had long been in demand for commercial uses, and while being transported from mines in France and Switzerland particles falling from the waggons were crushed under the wheels, their compression being aided by the heat of the sun, thus forming a very good road surface, which suggests the idea of trying it in street paving. The first experiment, which was on a macadam road, was encouraging, and in 1854 a portion of a street* in Paris was paved with asphalt on a concrete foundation. It met expectation, and four years later other streets were similarly paved. It grew rapidly in favor, and in 1869 asphalt pavements were introduced in London†; but not until 1880 were they given a trial in Berlin."

This pavement is now very extensively used in England and in Continental cities for paving both streets and footways.

SELECTION OF LINE AND LEVELS OF A NEW ROAD.

The following observations will refer principally to the construction of common roads through country where none previously existed.

From an engineering standpoint, the main object to be kept in view in the determination of the *route* and *levels* of a proposed line of communication, such as a new road, is to convey the traffic with a minimum expenditure of motive power, consistent, of course, with reasonable economy in construction. The points, therefore, to be secured, so far as the circumstances of the case will admit, are directness of line, including minimum length of route, easy curves, low summit levels, and easy inclinations or gradients.

To enable an engineer to select the most advantageous line for any proposed works, it is essential that certain data be first obtained by him as to the physical features of the surface of the country to be traversed, including the disposi-

---

* The *Rue Bergère*, laid with Val de Travers asphalt.
† When Threadneedle-street, near Finch-lane, was paved (May, 1869), by the Val de Travers Asphalt Company.

tion of the hills, valleys, rivers, &c., in the district, all of which information may be conveniently considered under the heading of *Engineering Geodesy*, the general order of the operations of which is usually as follows*:—

1. The *reconnaissance* or *exploration* of the country, to ascertain generally the facilities afforded for the proposed works, also to approximately determine the best *route*, noting the geological structure of the ground, the sources from which ma'erials may be obtained, and inquiring as to the trade, population and traffic of neighbouring towns. The engineer should also possess himself of the best plan of the district that can be obtained accurately showing the main features of the country.†

2. *Flying Levels* are often taken for ascertaining the elevations of detached points of importance, as regards the practicability and cost of the work and the selection of the line, such as passes across ridges and valleys, and points where structures of magnitude may be required.

A more definite selection of the line may now be determined by making

3. *Preliminary Trial Sections*, by taking continuous lines of levels, measuring both horizontal distances and heights, thus enabling the line to be set down with a degree of precision proportionate to the accuracy of the preliminary operations and the extent of ground to be included in the detailed survey to be fixed upon.

4. *Detailed Survey and Plan.*—The labour required on this will vary with the precision of previous operations. A suitable scale will be that of the Ordnance "parish maps"—*viz.*, $\frac{1}{2500}$ = 25·344 in. to a mile = 208·33 ft. to an inch. The plans should be shaded or marked with figured *contour*

---

* *See* Prof. Rankine's "Civil Engineering." Although some of the points mentioned in the following general order of operations may apply more particularly to railway or canal work, it will probably be instructive to give the *entire outline* of the methods of procedure on the part of the engineer in the design of lines of communication generally, as well as for *roads* in particular.

† The maps published by the Ordnance Survey Department are generally used whenever available, and the *6 in. to the mile* county map will probably be sufficient for a preliminary exploration.

*lines** to exhibit the undulations of high ground and valleys, and should show all "ridge lines," "valley lines," "passes," watercourses, buildings, orchards, ponds, or other conspicuous objects. A scale of distances may be marked on the plan along the *centre line* of the proposed road showing each *mile* and *furlong* from its commencement, also *bench marks* distributed along the line of work at distances of at least half a mile and near proposed structures of importance, as bridges, &c. It is also convenient to have a copy of the plans and section of the works on which to mark the results of trial pits and borings, and the estimated cost of each part of the work shown opposite its position on paper. In regard to the detailed survey and

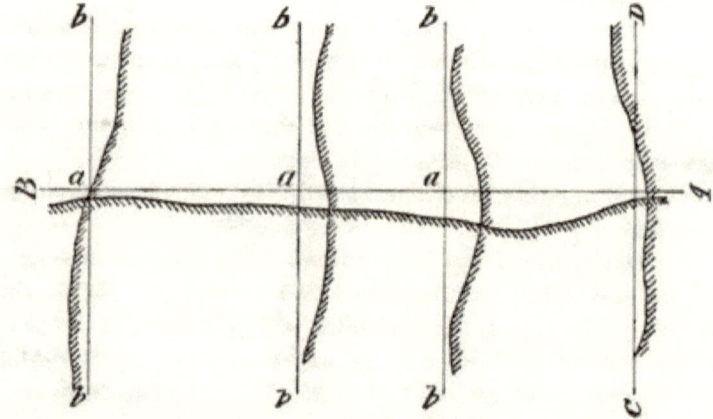

FIG. 5.—DIAGRAM ILLUSTRATING TREDGOLD'S METHOD OF SURVEYING BY RECTANGULAR LINES,

and showing Longitudinal and Transverse Sections along the main or centre lines.

plans, Mr. Tredgold, in his work on "Railroads," recommends that the survey be made "by rectangular lines, as infinitely superior to surveying by triangles," in giving an exact knowledge of the surface of the country. The accompanying diagram (Fig. 5), showing longitudinal and transverse sec-

---

* The exhibition on plans of the levels of the ground by means of *contour lines* is a very useful method, especially for the selection of lines of communication, drainage of towns, water supply, irrigation and drainage of lands, &c.

tions, is given to illustrate the method. "Let AB be a portion of the intended line, and CD the breadth of the country to be included in the survey. At any suitable distances choose stations, a, a, a, their distances apart depending on the changes of level, and let the *principal line AB*, and also the *cross lines bb, bb*, &c., be accurately levelled, and then *drawn*, as shown in the figure, *on the plan of the line of road*. If the distance *bb* is required to be considerable, perhaps an additional line in the principal direction may be necessary. *The etched lines show the form of the surface at the lines AB, bb, bb, &c., on the plan;* and the latter being *sections* at right angles to AB, there is no difficulty in seeing the extent of cutting or of embankment that may be avoided by varying the position of the principal line. In fact, *a plan of this kind*, to a person familiar with sections, *is better than a model of the country*."

5. *Additional Trial Sections*, longitudinal and transverse, are now made with the aid of the above detailed plan, in order to determine *exactly* and finally the best line obtainable.

6. *Marking the Line.*—The line being laid down upon plan is now to be staked out upon the ground by driving in, at distances usually a *chain* (66 ft.) apart, wooden stakes, 18 in. in length, along the centre line of the road.

7. *The Detailed Section* can now be prepared by accurate levellings over the line marked out. The section should show a horizontal *datum line*,[*] a line representing the natural surface of the ground, and lines showing the "*formation level*" and "*finished surface*" of the proposed works. Also figures should be given on the datum line for the *horizontal distances*, corresponding exactly with those on the plan, and the *heights* of the surface of the ground above datum require to be marked at each chain's length, as well as the depths of *cuttings* or heights of *embankments* to the formation level. The heights of the formation level and the *rates of inclination* must be

---

[*] A *datum line*=an imaginary line parallel with the horizon and with the lines of collimation. The "*Ordnance datum*"= the approximate mean water level at Liverpool. "*Trinity high-water mark*" is 12˙45 ft. above Ordnance datum.

figured on at each change of gradient; also any alteration in level necessitated to existing lines of communication, the heights of greatest floods, and the sectional area and velocity of rivers crossed, all require to be clearly shown.

8. *Trial Pits and Borings*, to ascertain the nature of the strata of the ground traversed, will be proceeded with simultaneously with operation No. 7, and the actual results observed should be marked on a plan, as suggested under heading No. 4. *Borings* are less costly, but *pits* are more satisfactory. The borings usually consist of vertical holes, about 4 in. in diameter, in the ground, and from which portions of the materials passed through are brought up, but are, of course, unfortunately reduced to a powder or paste by the action of the boring tool, thus necessitating trial shafts or pits for more complete information of the strata at important cuttings.

9. *Designs and Estimates.*—The foregoing information having all been duly obtained and recorded, the engineer will now be in a position to *design* the proposed works with such sufficient precision as will enable him to *estimate* the probable cost of same. The estimated cost of each part of the works may with advantage be entered upon a plan, as suggested under heading No. 4.

10. *Parliamentary Proceedings.*—If the works are such as will necessitate the application for an Act of Parliament to empower their execution, a *plan* and *section* and *book of reference* to the plans must be prepared and copies deposited as required by the standing orders of the Houses of Lords and Commons.

11. *Improving Lines and Levels under Powers of Deviation.*— The plans and sections prepared for obtaining the authority of Parliament are usually somewhat hastily made, and afterwards oftentimes require amending in respect to the precise lines and levels; therefore certain *powers of deviation* from the Parliamentary plan are taken to afford opportunity for such amendment as may afterwards be found necessary. The extent of these powers of deviation is usually as follows:—

Limits of Lateral Deviation.— In towns, 10 yards each side of centre line; in the country, 100 yards. This deviation is indicated on the Parliamentary plans by dotted lines.

Deviations of Level.—In towns, 2 ft.; in the country, 5 ft. Greater or less powers are conferred in special cases.

12. *Survey for Land Plans*—*i.e.*, plans to be used for the purchase of land and execution of the work, and surveyed carefully for plotting to a larger scale than was necessary under operation No. 4 above. If the *centre line* (whether straight or curved) of the proposed works has been carefully ranged and staked out, it may be utilised as a base for the secondary triangulation in this survey, and each stake used as a "station."

13. *Ranging and Setting-out the Line*—*i.e.*, marking the centre line of the work, as finally fixed upon, on the ground by means of wooden stakes.

14. *The Working Sections* are made from exact levels taken along the line finally settled upon and staked out. These sections should contain similar information to those mentioned under heading No. 7, but with greater precision, and be plotted to a larger scale. In adjusting the line of the formation level of the proposed works there should be an endeavour to equalise the earthwork—*i.e.*, to arrange that the earth from the cuttings shall equal, or only *slightly exceed*,* the quantity required for the embankments, and thus avoiding the expense of either obtaining more earth from "*side cuttings*" or of disposing of any surplus by forming it into "*spoil banks.*"

15. *Setting-out the Breadths of Land* required for the works on the land plans and also on the ground is the next operation. On the ground the *half-breadths* of the work, and the *total half-breadths*, as calculated, are laid off at right angles to the centre line, and the ends marked with stakes, and sometimes lines are also nicked out between the stakes showing the boundaries. A temporary post and rail *fence* of

---

* Newly-formed embankments always experience a settlement, more or less, and will therefore require to be made up to a proper level, requiring more earth.

larch or oak is then erected, before the earthwork is commenced, enclosing all the ground required for the work, usually including a narrow strip of land beyond the outer edge of the earthwork. The execution of the works is now proceeded with.

THE BEARING OF GEOLOGY UPON ENGINEERING WORKS.

In any large undertakings, such as the construction of roads, railways, canals, reservoirs, drainage, water supply, &c., which may involve deep and extensive "*cuttings*," the engineer is brought into close practical contact with the subject of *Geology*—a science which treats of the structure and constitution of the earth's crust, and of its history as regards rocks, mountains, minerals, rivers, valleys and other prominent physical features of its surface; so that to him some knowledge of this important subject must be of the greatest service, affording, as it does, a considerable amount of valuable information as to the nature and relative position of the various strata of the earth through or over which his works may have to cut or pass.

Examples of almost all known geological formations are to be found in Great Britain, and, so far as is known, it furnishes a history more complete than that of any other tract of similar extent. A good general knowledge of the location of the various formations and the "*outcrop*" of any particular stratum may be readily obtained from a study of a coloured *geological map* with sections of the country, and displaying the superficial exposures of the various strata, such, for example, as that issued by the Geological Survey.

The *general geological structure of Great Britain* may be learnt from a careful examination of the map, and it will be observed that the MESOZOIC and CAINOZOIC rocks occupy the middle, eastern and south-eastern portions of England; while the older, or PALÆOZOIC, rocks occupy Cornwall and Devon, Wales, North of England and Scotland. The TERTIARY rocks, it will be seen, lie principally in two districts, called the London basin and the Hampshire basin respectively. When the Tertiary clays and sands of the London basin are penetrated by borings or wells the *chalk* is reached, and beyond the limits of the Tertiary strata the members of the CRETACEOUS strata come to the surface in succession.

Tracing the cretaceous formations towards the west, we at length find the OOLITE coming up from under them; the members of the Oolite having, like the Cretaceous strata, a general "*dip*" to the east. Passing across the "*outcrop*" of the Oolitic series as they come to the surface, one after another, we come to the LIAS; the Oolite in many places along the boundary forming an "*escarpment*" which overlooks the Lias, plainly showing that at one time the Oolite extended farther to the west than it does now. The chalk, however, extended even farther than the Oolite, and an isolated patch will be seen on the map, which rests directly on the Trias and Palæozoic rocks in Devonshire. Still farther to the west we reach the TRIAS, coming up from under the Lias. Proceeding farther, we come upon the PALÆOZOIC rocks, which cover at least two-thirds of the whole island. Being more disturbed than the newer rocks, the Palæozoic strata have not the same uniformity of dip as the former; but still the general dip is to the east, and we find the lowest members, for the most part, in the extreme west and north—the oldest known rocks, the LAURENTIAN, occurring in the Hebrides and north-western extremity of Scotland.

"The Palæozoic strata are usually highly indurated, often metamorphosed, and associated to a great extent with igneous rocks, and, having suffered much by denudation on account of their great age, confer upon a great portion of the island a rugged and mountainous character."*

The *order of succession* of the different strata composing the crust of the earth, and the principal sub-divisions of geological time, are set out in the following :—

---

* "Geology," by W. S. Davis.

MESOZOIC OR SECONDARY. (AGE OF REPTILES.)	OOLITE		{ Upper Middle Lower }
	LIAS	… … …	
	TRIAS	… … …	
	PERMIAN	… … …	
PALÆOZOIC OR PRIMARY. (AGE OF INVERTEBRATES.) (AGE OF FISHES.) (AGE OF ACROGENS.)	CARBONIFEROUS	… …	
	DEVONIAN AND OLD RED SANDSTONE	… … …	
	SILURIAN	…	{ Upper Lower }
	CAMBRIAN	… … …	
	LAURENTIAN	… … …	

Upper
　　Middle　⎫
　　Lower　⎪
　　　　　　⎪
　　ay　　　⎪
　　　　　　⎪
　　　　　　⎬ Jurassic.
　　　　　　⎪
　　)olite　⎪
　　e　　　 ⎪
　　　　　　⎭

　　y and Sand　⎫
　　　　　　　 ⎪
　　y and Limestone ⎫
　　rth Beds　　　　⎪
　　ed Marl)　　　　⎬ Poikilitic (variegated).
　　ed Sandstone)　 ⎪
　　estone　　　　　⎭

　　ie Shale.
　　.imestone.
　　ie Shale.

　　ery Beds.
　　ry Beds.
　　oc Beds.

　　s.

ındstones and slates (older than the Silurian) occurring
n North Wales.
ancient name of Wales.

ighly contorted gneissic rocks, being the oldest known
cks in the world.

GENERAL OBSERVATIONS AS TO ROUTE OF LINE OF ROADWAY, INCLINATION, &c.

The primary object for the construction of a line of roadway through a new district is usually for the connection of two or more distant towns with each other, or, probably, with the sea coast or a port. In the design of such a line of communication a first consideration would be to make the road as short as possible between the various points to be connected; the endeavour obviously being, therefore, as far as possible to preserve a perfectly *straight line* from point to point. Straightness, however, has often to be sacrificed in order to secure good gradients and to keep down the cost of construction, which would otherwise be greatly enhanced by passing through deep cuttings or by building heavy embankments. Although the length of the circuitous route may exceed that of the direct line, yet if the inclinations of the former are less than those of the latter, with the same expenditure of horse-power the loss through increase of distance may be exceeded by the gain in speed due to the flatter gradients. Suppose, for example, one road has an inclination of 1 in 20, and another inclines at the rate of 1 in 40, then the same power will propel (at the same rate) a load through 20 ft. of the one and through 40 ft. of the other. The advisability of so winding the route of a road as to avoid crossing elevations in the direct line, and thus probably save many feet of perpendicular height, will therefore depend upon the length by which the road will be increased by so doing.

In laying down a new line across undulating country the relative advantages and disadvantages of the various curves and inclinations must be considered, with a view to ascertaining what rate of inclination may be adopted for a given decrease in length. This will depend upon the amount of resistance offered to traction upon the road.

The *cost* of any undertaking will necessarily depend largely upon the uniformity or otherwise of the surface of the country traversed, upon the number of rivers and watercourses or lines of communication to be crossed by means of bridges or tunnels, upon the nature of the strata, and also upon the amount of deviation *en route* necessary either to avoid tunnelling or cuttings of a deep and expensive nature, or to serve neighbouring towns or villages.

In carrying a line of road across a deep valley it should generally cross as high up the valley as practicable, where the descent and ascent will be less. A narrow part of the valley should also be chosen, and an endeavour should be made to find firm ground for the foundation of the embankment or viaduct, and the crossing should be as nearly at right ang'es to the direction of the valley as possible. This latter principle applies to bridges over rivers, or over or under other lines of communication. According to Prof. Rankine, "the cost of a skew bridge increases nearly as the square of the secant of the obliquity."

If the sides of a valley are considerably contracted at some point lower down the stream, or if there should be at such a point isolated hills standing in the valley, it may then require less embankment to bring the road to the necessary height, and thus be advantageous for the line of road to cross here. Accurate cross sections would therefore require to be taken through the valley at various likely points, in order to satisfactorily ascertain the most economical position to cross.

In passing deep valleys or ravines in mountainous countries high arches of masonry may be found necessary. Such were adopted by Mr. Telford in some parts of Scotland—one of the most important erected by him being the bridge over the Mouse Water at Cartland Craigs, on the Lanark road. This viaduct consists of three arches, each of 50 ft. span, the soffites of which are 120 ft. above the surface of the water below.

In crossing a "*ridge*" advantage may generally be taken of the lowest "*pass*" for so doing, if conveniently accessible. The crossing should be made as nearly at right angles as possible. "When a line of communication runs along one side of a valley the obstacles which it has to cross are chiefly the small branch valleys that run into the main valley, and the promontories or ends of branch ridge lines that jut out into the main valley between the branch valleys. In this case the greatest economy of works is attained *by taking a serpentine course,* concave towards the main valley in crossing the branch valleys and convex towards the main valley in going round the promontories, except where narrow necks in the promontories and narrow gorges in the branch valleys enable a more direct course to be taken with a moderate amount of work."*

In settling upon the *rates of inclination* in the design of a new road it is desirable to first fix upon a *ruling or maximum gradient.* An inclination which should not be exceeded, if it is practicable to avoid so doing, is that which will present an impediment to fast driving either in ascending hills or in descending. A gradient of 1 in 35 has been found to perfectly fulfil these conditions, and was adopted by Mr. Telford on a part of the Holyhead road on the north of the city of Coventry. "It may, therefore, be taken as a *general rule,* in laying-out a line of new road, never, if possible, to have a greater inclination than that of 1 in 35."†

"The great fault of most roads in hilly countries is that, after ascending a considerable height, they constantly descend again before they gain the summit of the country which they have to traverse. In this way the number of feet actually ascended is made many times more than would be the case were no height, once gained, lost again."† Instances of this defect were given in the course of the description of "Telford's Roads."

"A perfectly *flat road* is to be avoided, if it is not to be raised by embanking at least 3 ft. or 4 ft. above the general level of the land on each side of it, so as to expose the surface of it fully to the sun and wind; for if there is not a longitudinal inclination of at least 1 in 100 on a road water will not run off, in consequence of which the surface, by being

---

* Prof. Rankine's "Civil Engineering."
† "A Treatise on Roads," by Sir H. Parnell.

for a longer time wet and damp than it otherwise would be, will wear rapidly away, and the expense of maintaining it in order by scraping and laying on materials will be very much increased."* So that a gradient of from 1 in 100 to 1 in 150 may be taken as the *flattest* that is desirable for the easy drainage and durability of the road; also, these moderate inclinations are probably favourable to the ease of draught by horses.

A road should be well exposed to the joint action of the sun and wind, in order to keep it as *dry* as possible, as even the hardest road materials wear much more rapidly when wet and damp. It will be advantageous, therefore, in passing through undulating country, that the road should be on the north sides of the valleys traversed, and all high fences, walls, banks or overhanging trees, excluding the free action of the sun and wind, and thus not permitting the evaporation of moisture, are detrimental to the durability of the road.

The facilities offered by any particular district for cheaply obtaining road materials may in some cases afford good reasons for deviating the course of a proposed road from the direct line.

RIVERS AFFECTING ROUTE.

"The peculiar circumstances of a river may render it necessary to deviate from a direct line in laying-out a road.

"A difficulty may arise from the breadth of the river requiring a bridge of extraordinary dimensions, or from the land for a considerable distance on the side of the river being subject to be covered with water to the depth of several feet in floods.

"In these cases it may appear, upon accurately calculating and balancing the relative inconvenience and expense of endeavouring to keep a straight line and of taking a circuitous route, that upon principles of security, convenience and expense the circuitous course will be the best.

"In general, rivers have been allowed to divert the direct

---

* "A Treatise on Roads," by Sir H. Parnell.

line of a road too readily. There has been too much timidity about incurring the expense of new bridges, and about making embankments over flat land to raise the roads above the level of high floods.

"These apprehensions would frequently be laid aside if proper opinions were formed of the advantages that arise from making roads, in the first instance, in the *shortest directions* and in the most perfect manner. If a mile, half a mile, or even a quarter of a mile, of road be saved by expending even several thousand pounds, the good done extends to posterity, and the saving in annual repairs and horse labour that will be the result will before long pay off the original cost of the improvement."*

CROSSING EXTENSIVE PLAINS, MARSH GROUND, &c.

In order to keep a roadway or other line of communication dry, it is usually necessary to keep its surface well above the general level of the surrounding country when carrying it across an extensive plain, especially if the district traversed is subject to inundations. In embanking for this purpose trenches A, A (Fig. 6), may be dug on each side of the line of

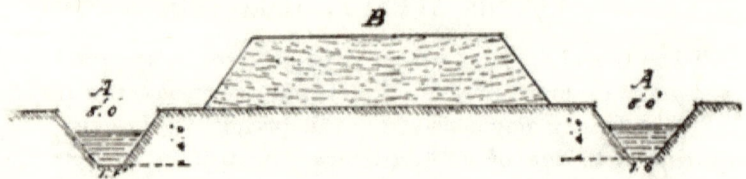

FIG. 6.—METHOD OF EMBANKING WHEN LINE CROSSES TRACT OF FLAT COUNTRY SUBJECT TO FLOODS.

road and the excavated material used in the embankment B. The trenches also serve to collect the surface water, which may be discharged into the nearest watercourse.

One of the most difficult and expensive obstacles to meet and overcome in the line of a new road is that presented by *boggy and marsh ground*, and in such cases it will generally

---

* "A Treatise on Roads," by Sir H. Parnell.

be found to be the most satisfactory course to deviate from the direct line if this kind of soil can thus be avoided without largely increasing the length of the road. If, however, there is no reasonable alternative to the crossing to the marsh, it will be necessary to properly drain the subsoil, to consolidate it, and to construct a high embankment.

The following is a specification drawn up by Mr. Telford, and successfully acted upon in forming a road over a peat bog in Ireland:—

"When the line of the road has been traced out to the exact width and line of direction, main drains are to be cut on each side 8 ft. wide at top, 4 ft. deep and 18 in. wide at bottom. The peat dug out of these drains is to be spread over the surface of the roadway in form of a ridge, taking care to previously cover all the very soft and swampy places with dried peat, sods or brushwood. Numerous drains are to be cut across the roadway from the one main drain to the other; they are to be 3 ft. deep at the centre of the roadway and 4 ft. deep at the main drains. After the whole have remained in this state for two summer months the bed for the roadway is to be neatly formed, with the sides on the same level, and with a convexity of half an inch in the yard.

"The carriageway is then to be covered with 6 in. of clay, laid on evenly, and firmly compressed by stampers or rollers; it is to have a fall of 1 in. in the yard from the centre towards the sides. Over the clay is to be put 4 in. of small gravel; it is to be frequently rolled, and when solid and compressed the *foundation* will be formed for the reception of the road materials."

Should the ground to be crossed by a line of communication be so soft that an ordinary embankment would sink in it, any of the following expedients,* given in the order of increasing difficulty, may be adopted as the case may require:—

1, Increase the firmness of the natural ground by digging side drains.

2, Give the sides of the embankment the same slope as the

---

* Suggested by Prof. Rankine in his "Manual of Civil Engineering."

"angle of repose" of the material of the natural ground, and thereby broaden its foundation.

3, Dig a trench and fill in with solid material, as in sketch (Fig. 7).

4, Compress and consolidate the ground by means of short piles.

5, Make the embankment of light materials, forming a sort of raft floating on the soft ground, as hurdles, fascines or dry peat. George Stephenson, in carrying the Liverpool and Manchester Railway across Chat Moss, used dry peat to form the embankment, and upon this he placed a platform of two layers of hurdles to carry the ballast.

6, Throw in stones, gravel and sand until an embankment

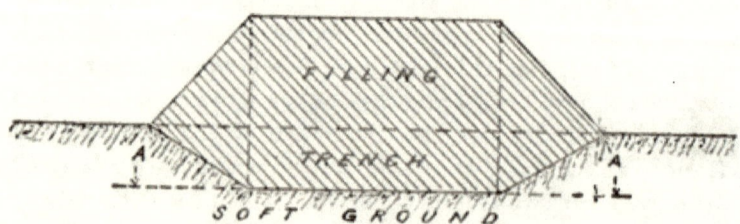

FIG. 7.—METHOD OF EMBANKING THROUGH SOFT GROUND.
A = "angle of repose" of the soft material.

is formed, resting on the hard stratum below. The material thrown in takes up the same natural slope as if in air.

Bog Roads, Ireland.—These roads present frequent difficulty to roadmakers in Ireland—a great portion of the central plain of that island being covered with bog land, which occupies about two-fifths of its entire surface. The surveyor of the county of Armagh states[*] that "roads constructed over peat bogs are very liable to get out of shape, owing either to the movement of the bog, to one or both sides being cut away, or to the traffic. In the latter case, owing to the bulk of the traffic passing over the middle of the road, the centre becomes depressed and the sides rise. No remedy appears to

---

[*] "Proceedings of the Association of Municipal and County Engineers," vol. xviii.

have been found yet to prevent this distortion of bog roads .... In making *new roads over bogs* the method usually adopted is to cut deep drains on each side of the proposed roadway and some 10 ft. back from the line of fences, the stuff taken out being spread over floor of proposed roadway; small transverse cuts are also made at frequent intervals, and field pipes laid down and covered over. When the bog has been well drained it should be brought to a fair level and covered with layers of sods, until formation surface of proposed roadway is at least 2 ft. over flood level; on this surface should be spread a layer of whins, then a coating of clay loamy gravel, 9 in. thick, and when this has consolidated the top metal can be spread off. It is useless to put a *pitched foundation* in a bog road, as the stones lose their position in a short time, and either rise to the surface or get lost in the bog."

EARTHWORK, EMBANKMENTS, CUTTINGS AND THEIR DRAINAGE.

With a view to economy of construction, the longitudinal inclinations of a road should be laid out with a minimum of "*cutting*" and "*embanking*"; which line is to be obtained by a judicious adaptation of the facilities offered by the undulations of the country, and by calculations of the quantities of earth to be dealt with. It is also important that, where a road must be carried over a high elevation, it should be so laid out "that it shall not have any fall in it from the point from which it departs till it reaches the summit. The lowering of heights, and the filling of hollows, should be so adjusted as to secure gradual and continued ascending inclinations to the highest point to be passed over."*

In order to ascertain the nature of the ground to be excavated previous to the execution of a piece of earthwork, *trial shafts* should be made at important or deep cuttings, and, at intervals of about 200 yards, *borings* may be made sufficient to indicate any alteration in the strata necessitating a shaft being sunk to obtain more accurate information. A

---

* Sir H. Parnell, "Treatise on Roads."

knowledge of the nature of the ground is necessary to determine the safe *slopes* of the sides of the proposed cuttings, or of embankments to be made with the material. It will be convenient to have the earthworks numbered on the plans or sections, giving opposite them the cubic contents of each and the intended slopes of the sides The working plan should also show the position of the stakes along the centre line, the width of the road, with outer lines showing the extent of the slopes. The stakes must be numbered, and the widths or half-widths of the land required (including that occupied by the roadway, slopes, and generally an additional 6 ft. on each side for a bank and ditch) can be set out upon the ground at each chain length, stakes also being left to mark the outer edge of the slopes. The intended levels of the formation surface should be set up by fixing stont posts along the centre line, marking the points of variation in inclinations and any intermediate levels necessary, so that the line may be "boned" through in order to obtain the required gradient. In practice the workmen usually hold up a boning staff of a known length between certain accurately-fixed levels (previously given them by the person in charge of the works), so that when its top edge is lineable with these levels its foot indicates the required height of the intermediate point.

Embankments are formed either *in one layer*, by discharging earth-waggons over the sloping extremity of the bank known as the "*tip*," or *in two or more thick layers* each by tipping over the end as above, but giving an interval for settlement before commencing the second layer, or, in the third instance, by spreading the earth *in thin layers* up to 18 in. thick and consolidating it by ramming. The first method is the quickest, and is generally adopted for ordinary embankments, excepting, of course, embankments forming dams in connection with water supply reservoirs, &c., which require special treatment and additional care in order to be watertight and to resist the lateral pressure to which they will be subjected.

Embankments require to be made with care, and to prevent future slipping the base must be formed at first to its full breadth. "The earth should be laid on in concave courses (see Fig. 8), in order to give firmness and stability to the work. It is not at all uncommon in many parts of the

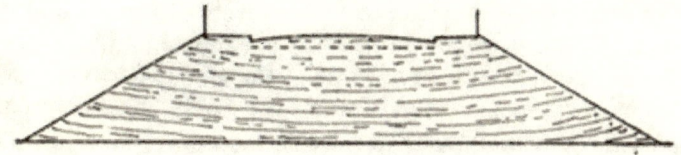

FIG. 8.—EMBANKMENT FORMED CONCAVELY.

country to see embankments formed convexly (Fig. 9), the consequence of which is that they are for ever slipping."*

In the method of forming embankments in one layer by embanking out from one end, it is "best to form the outsides of the embankment first and to gradually fill in towards the centre, in order that the earth may arrange itself in layers with a dip from the sides inwards; this will in a great measure counteract any tendency to slip outwards. The foot of the slopes should be secured by buttressing them either by a low stone wall, or by forming a slight excavation for the same purpose."†

For increased safety the slopes of an embankment may be of a greater inclination than the natural slope of the earth;

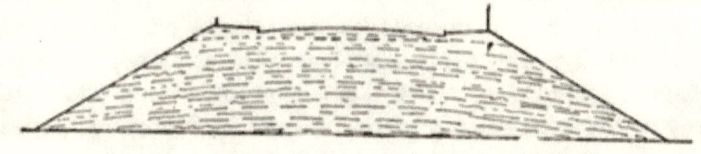

FIG. 9.—EMBANKMENT FORMED CONVEXLY.

and the surface water falling upon the top should not be allowed to run down the slopes and so form channels.

Efficient drainage is necessary to prevent *slips* in cuttings,

---

* Sir H. Parnell, "Treatise on Roads."
† D. H. Mahan, "A Treatise on Civil Engineering."

but, as an additional precaution, the slope may be covered with a closely-packed layer of stones, or a dry stone, brick, or masonry retaining wall may be built at the foot of the cutting.

The green sod and fertile soil taken from the surface of the land cut through should be carefully collected and retained, in order that it may be relaid upon the slopes when formed; or, in the absence of this, the slopes can be covered with 4 in. of surface mould and sown with hay seeds, so that they may soon be covered with grass, which will tend largely to prevent slipping. If stones are easily obtainable, a wall about 3 ft. high may be built to support the foot of the slopes and prevent their falling into the side channels.

" Where *slips* occur from the action of *springs*, it frequently becomes a very difficult task to secure the side slopes. If the sources can be easily reached by excavating into the side slopes, drains formed of layers of fascines or brushwood may be placed, to give an outlet to the water and prevent its action upon the side slopes. The fascines may be covered on top with good sods laid with the grass side beneath, and the excavation made to place the drain be filled in with good earth, well rammed. Drains formed of broken stone, covered in like manner on top with a layer of sod to prevent the drain from becoming choked with earth, may be used under the same circumstances as fascine drains. Where the sources are not isolated, and the whole mass of the soil forming the side slopes appears saturated, the drainage may be effected by excavating trenches a few feet wide at intervals to the depth of some feet into the side slopes and filling them with broken stone, or else a general drain of broken stone may be made throughout the whole extent of the side slope by excavating into it. When this is deemed necessary, it will be well to arrange the drain like an inclined retaining wall, with buttresses at intervals projecting into the earth further than the general mass of the drain. The front face of the drain should, in this case, also be covered with a layer of sods with the grass side beneath, and upon this a layer of good earth should be compactly laid to form the face of the side slopes.

The drain need only be carried high enough above the foot of the side slope to tap all the sources, and it should be sunk sufficiently below the roadway surface to give it a secure footing."*

The *stability* of earthwork depends upon the resistance offered by its parts to slipping on each other, from the friction between the grains and from their mutual adhesion; which latter is reduced by the action of frost and thaw, by wetness, variation of weather, &c. The *permanent stability*, therefore, depending entirely upon friction, whilst adhesion affords an additional stability of a temporary nature, which is useful in enabling new cuttings to stand with vertical faces for a limited time and depth below the surface, depending upon the class of earth cut into.

The angle of inclination to the horizon at which the side of a cutting or embankment will stand permanently stable is known as the " *natural slope*," or " *angle of repose* " of the material forming the slope.

Sir Henry Parnell, in his important " Treatise upon Roads," states that " the slopes of the banks of a deep cutting should never be less, except in passing through stone, than 2 ft. horizontal to 1 ft. perpendicular." The inclinations most generally adopted are $1\frac{1}{2}$ to 1 and 2 to 1; steeper slopes exclude the sun and wind. Also, the slopes at which cuttings and embankments can be safely made depends entirely upon the nature of the soils, for example:—

	Horizontal.		Perpendicular.
London and plastic clay, maximum slope	3	to	1
Chalk or chalk marl, „ „	1	to	1
Sandstone (if solid, hard and uniform), maximum slope	$\frac{1}{4}$	to	1
Oxford clay, maximum slope	2	to	1
Limestone strata (if solid), maximum slope	$\frac{1}{4}$	to	1
Limestone strata (mixed with clay beds), maximum slope	$1\frac{1}{2}$ or 2	to	1
Granite, slate and gneiss, maximum slope	$\frac{1}{4}$	to	1

---

* D. H. Mahan, " A Treatise on Civil Engineering."

The *natural slopes of earths* (with a horizontal line) as given by Molesworth and Rankine are as follows :—

	Molesworth Degrees.	Rankine Degrees.
Gravel, average...	40	48 to 35
Dry sand	38	37 to 21
Sand	22	—
Vegetable earth	28	45 to 14 (Peat)
Compact earth ...	50	—
Shingle ...	39	48 to 35
Rubble ...	45	—
Clay, well drained	45	45 (Damp clay)
Clay, wet	16	17 to 14

The following is a table of the *Lengths and Angles of Slopes**:—

Slope.	Angle with Horizon.	Length (Height taken as 1·00).
¼ to 1	75° 58′	1·0307
½ to 1	63° 26′	1·118
¾ to 1	53° 8′	1·25
1 to 1	45° 0′	1·4142
1¼ to 1	38° 40′	1·6
1½ to 1	33° 42′	1·802
1¾ to 1	29° 44′	2·016
2 to 1	26° 34′	2·236
3 to 1	18° 26′	3·162
4 to 1	14° 2′	4·124

Fresh-made embankments always subside to an amount depending upon the class of material of which they are formed, the settlement usually varying from one-twelfth to one-fifth † of the original height.

For economy of execution, the line of *formation level* of a proposed work requires to be so adjusted that the earth derived from the *cuttings* shall be equal to that necessary for making the *embankments*. With practice this line may be laid down by the eye, sufficiently accurate for practical pur-

---

* Molesworth's "Engineers' Pocket-Book."
† Rankine's "Civil Engineering."

poses, upon a section of the ground. Should there be a surplus of cutting over embankment, it has usually to be disposed of in "*spoil banks*," or if there should be an excess of embankment over cutting the additional earth must be obtained by "*side cutting*"; thus, in both cases, necessitating expense for labour and land, but the distance of transmission of earth from various points along the line will affect the question of cost.

Embanking upon "Sidelong" Ground.—In "sideforming," or the making of embankments along the sides of hills, the slope to be covered should be cut into "steps" or "benched out," in order to ensure the stability of the superimposed

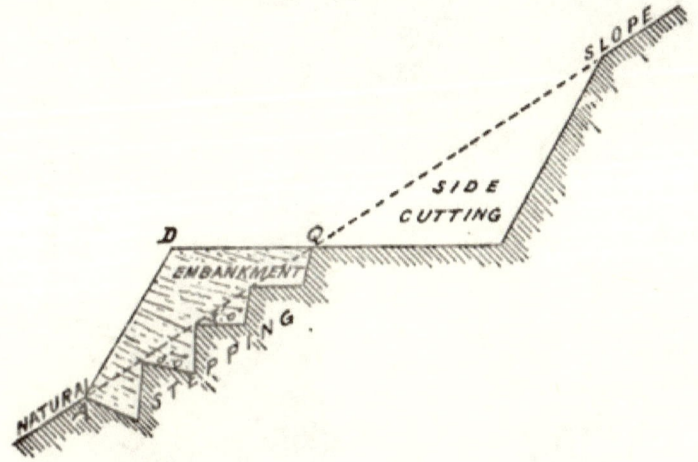

FIG. 10.—EMBANKING UPON "SIDELONG" GROUND.

earth and prevent slipping (*see* Fig. 10). It is also important to divert all land springs, and the earth should be well compressed. "The best position for the steps is perpendicular to the axis of greatest pressure . . . , so that, if A D is inclined at the angle of repose, the steps near A should be inclined to the horizon in the opposite direction to A D . . . , while the steps near Q may be level."*

Catch-Water Drains.—In the accompanying sketch (Fig. 11)

---

* Rankine's "Civil Engineering."

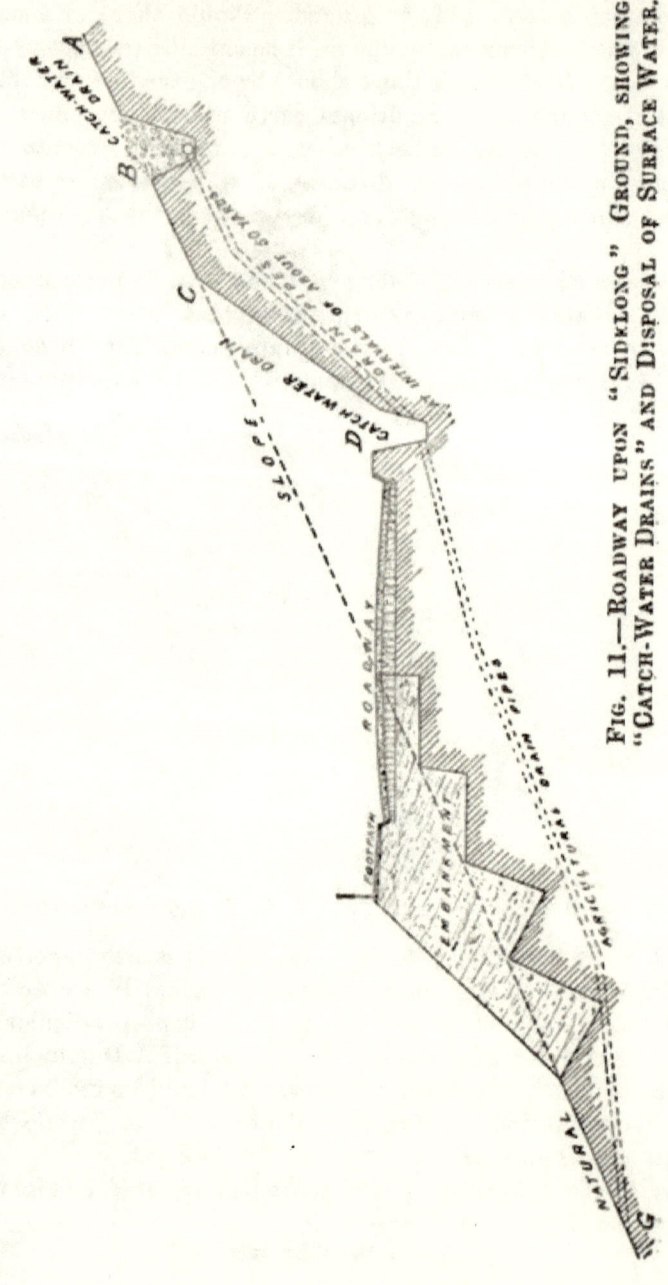

Fig. 11.—Roadway upon "Sidelong" Ground, showing "Catch-Water Drains" and Disposal of Surface Water.

B is a "catch-water drain" cut on the uphill side of a roadway made through "sidelong" ground. Its object is to intercept the surface water descending from high ground A towards C, and thus prevent damage which would otherwise be occasioned to the slope of the cutting. The water from B is conveyed by means of agricultural drain pipes to the catch-water trench D, which may, according to circumstances, either have an outfall into the nearest watercourse available, or its contents may be discharged through pipes from D to G as in sketch, the outlet G always being upon the natural ground.

Catch-water drains should, of course, be commenced at the outfall, and may consist simply of an open ditch about 4 ft. wide and 2 ft. or 3 ft. deep, or may be formed of earthenware agricultural pipes, laid in a trench and covered with broken stones, or brick or stone underground conduits may be constructed.

The safest materials for earthwork are those which will permit of water passing through them, such, for example, as gravel, clean sand and shivers of rock. Clay, when dry, is also safe, but mixtures of sand and clay are bad, the sand admitting water but the clay retaining it.

Retaining Walls are frequently required in forming a road through steep and hilly country. "If a retaining wall be built of brick, the thickness at top should be one brick, or 9 in., and it should increase in breadth by onsets of half a brick at every eight courses to the level of the road, below which the thickness for the stepping of the foundation should increase half a brick at every four courses to the bottom. All the walls of this description should batter in a curve line on the face at the rate of 1 in. in every foot "[*] (*see* Fig. 12).

Sir Benjamin Baker, "as a result of his own experience, makes the thickness of retaining walls in ground of an average character equal to one-third of the height from the top of the footings." Also, "a wall quarter of the height in thickness and battering 1 in. or 2 in. per foot on the face,

---

[*] Sir H. Parnell, "Treatise on Roads."

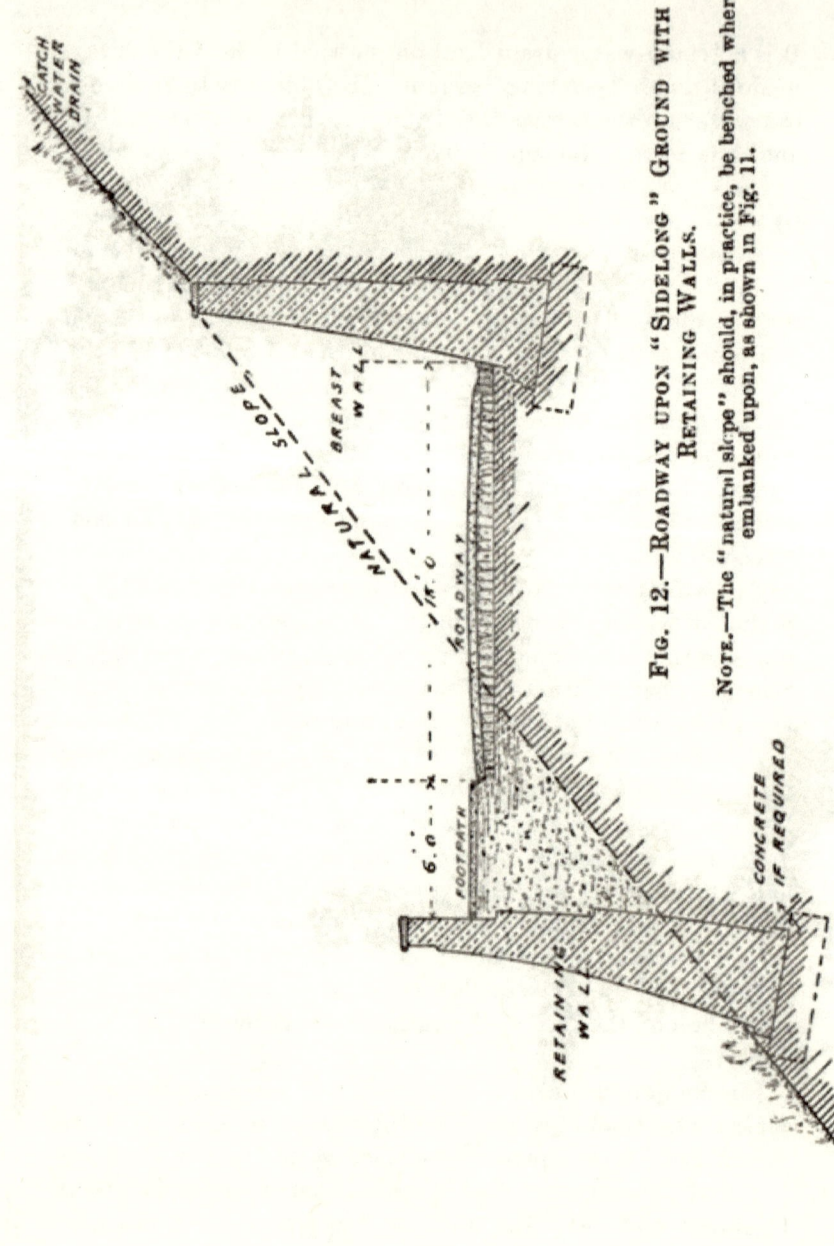

Fig. 12.—Roadway upon "Sidelong" Ground with Retaining Walls.

Note.—The "natural slope" should, in practice, be benched where embanked upon, as shown in Fig. 11.

possesses sufficient stability when the backing and foundation are both favourable." And "under no ordinary conditions of surcharge or heavy backing is it necessary to make a retaining wall on a solid foundation more than double the above or half of the height in thickness." He further states, as to the equivalent fluid pressure, that "experiment has shown the actual *lateral thrust* of good filling to be equivalent to that of a *fluid weighing about 10 lb. per cubic foot*, and allowing for variations in the ground, vibration and contingencies, a factor of safety of 2, the wall should be able to sustain at least 20 lb. fluid pressure, which will be the case if quarter of the height in thickness."

"The foundations of retaining walls are subject to considerable pressures, which, moreover, are not uniformly distributed. They should, therefore, be of such a width that the maximum intensity of pressure is not greater than the soil can safely bear, and the centre of pressure should not be nearer the outside edge than one-third the width of the foundation."

"Ordinary firm earth will safely bear a pressure of about 1 to 1½ ton per square foot, while moderately hard rock will bear as much as 9 tons."*

A road may be formed on the face of a precipice by building a heavy retaining wall, cutting into the rock, and making good the space behind the wall with filling material. If there is to be a footpath it should be upon the outer side next the retaining wall, so that the greater part of the weight of the heavy traffic will be by this means carried by the natural rock.

Breast Walls.—When the earth behind a wall is sufficiently consolidated to stand vertically alone the wall need only be thick enough to resist its own weight, as it then acts simply as a covering to the earth to prevent the degrading action of the weather. Walls of this class are known as "breast walls."

THE COMPUTATION OF QUANTITIES OF EARTHWORK.

The quantities of earthwork may be accurately calculated

---

* 'Notes on Building Construction," vol. iv.

by dividing the various solid masses into definite geometrical forms, such as pyramids, prismoids, wedges, &c., the cubic capacities of which are obtained by ordinary " mensuration."*

A more expeditious but less accurate mode is to divide any given length of cutting or embankment into a number of equi-distant perpendicular transverse sections, and then calculate the cubic capacity between each pair by multiplying the half sum of the pair of sectional areas by the distance between them, or, the area of a single middle section may be taken (assuming it to be a fair average section of the length to be calculated), and multiplying by the distance as above.

The method of taking the average of the top and bottom *breadths* and average *depths* at various points, and multiplying the product of these by the *length* of the work, is also sometimes adopted, but is likely to lead to erroneous results.

A more accurate method is to find the areas of a number of equi-distant perpendicular transverse sections,† and then proceed with them as if they were so many equi-distant ordinates, the result being the cubic capacity between the first and last cross sections.

The rule for calculating irregular figures by means of equi-distant ordinates is known as "*Simpson's Rule*," and is as follows:—

Divide the given length into any *even* number of *equal* parts, then "*add together the first ordinate, the last ordinate, twice the sum of all the other odd ordinates and four times the sum of all the even ordinates, multiply the result by one-third of the common distance between two adjacent ordinates.*"

To put this in another form—

Let A = sum of areas of the first and last ordinates (*i.e.*, of cross sections Nos. 1 and 5, Fig. 13).

Let B = sum of all other *odd* ordinates (cross section No. 3).

---

* See "Notes on Practical Sanitary Science," part ii., "On the Calculation of Areas and Cubic Space," by William H. Maxwell, published by the Sanitary Publishing Company, 5 Fetter-lane, E.C.

† The result will be more accurate the greater the number of sections taken, and the sections should be nearer together on uneven ground than on level.

E

Let C = sum of all the *even* ordinates (cross sections Nos. 2 and 4).

Let D = common distance between two adjacent ordinates. Then—

$$(A + 2B + 4C) \times \tfrac{1}{3}D = \text{cubic contents.}$$

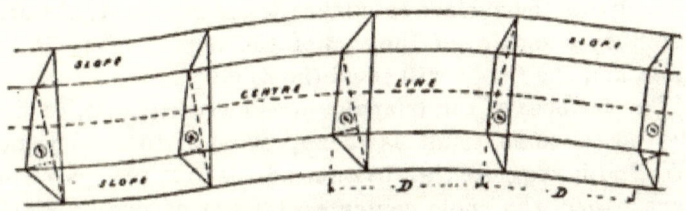

FIG. 13.—CALCULATION OF IRREGULAR FIGURE BY "SIMPSON'S RULE."

The above method, though very fairly correct if the cross sections are taken at *short* intervals, is a somewhat tedious one, and in practice would involve too much time, as all the sections require to be plotted to calculate their respective areas.

The *Prismoidal Formula* is frequently used, and is as follows: "The area of each end added to four times the middle area, and the sum multiplied by the length divided by 6, will give the solid content."

Or—

$$\frac{\left[\text{Sum of areas of both ends} + (\text{middle area} \times 4)\right] \times \text{length}}{6}$$

In practice all the above methods are generally superseded by the more rapid means of arriving at the contents by the use of *Earthwork Tables*, such, for example, as those compiled by Sir John Macneill,[*] which are based upon one form of the prismoidal formula. The tables give at sight, for various depths of cuttings or heights of embankments, the contents of 1 chain in length for different widths of, say, from 1 ft. up to 30 ft. or 40 ft., of the *central portion* of cutting, &c., in cubic

---

[*] "*Tables for Calculating the Cubic Quantity of Earthwork in the Cuttings and Embankments of Canals, Railways and Turnpike Roads*," by Sir John Macneill, C.E., F.R.A.S., &c.

yards; also the contents of 1 chain in length of *both slopes* in cubic yards for various slopes of, say, ¼ to 1 up to 2 to 1.

It may be noted that the cubic contents of the slopes increase proportionately to the square of the depth of the cuttings.

"To measure the solidity over large areas of irregular depth, divide the surface into triangles, and multiply the area of each by one-third of the sum of the depths taken at the angles, and the result will equal the solidity.

"The surfaces of the triangles must be true planes, or they must be taken so small as to approximate to true planes. J. W. Smith suggests the division of the area into parallelograms; then the cubic contents = ¼ (sum of depths at the angles) × (area of parallelogram)."*

In calculating the solid contents (of earthworks) allowance must be made for the difference in bulk between the different kinds of earth when occupying their natural bed and when made into embankment. From some careful experiments on this point made by Mr. Elwood Morris, published in *The Journal of the Franklin Institute*, it appears that *light sandy earth* occupies the same space both in excavation and embankment; *clayey earth* about one-tenth less in embankment than in its natural bed; *gravelly earth* about one-twelfth less; *rock*, in large fragments, about five-twelfths more, and in small fragments about six-tenths more."†

ROAD DRAINAGE.

Too much attention can scarcely be paid to the question of drainage in connection with the construction of roads. "Roads *kept dry*," says Sir Henry Parnell in his important "Treatise on Roads," "will be maintained in a good state with proportionally less expense. It has been well observed that the statuary cannot saw his marble, nor the lapidary cut his jewels, without the assistance of the powder of the specific materials on which he is acting; this, when combined

---

* "Molesworth's Engineers' Pocket-Book." (Hurst.).
† "The Construction of Roads and Streets," by H. Law and D. K. Clark.

E²

with water, produces sufficient attrition to accomplish his purpose. A similar effect is produced on roads, since the reduced particles of the materials, when wet, assist the wheels in rapidly grinding down the surface."

The contour of the transverse section of a road should be so formed that the whole of its surface drainage may be speedily carried away to the side channels and from thence into the side ditches, as illustrated in the cross section (Fig. 14). If, in a 30-ft. roadway, the centre 10 ft. of width be formed to a regular curve of 90 ft. radius, and the two sides or haunches with a maximum slope of 1 in 20 to the side channels, a self-draining cross section will then be obtained. For side slopes of 1 in 20 the height of the "crown" of the roadway above the level of the side channels will be 8 in.

In country roads an open main drain or ditch, discharging into the nearest natural watercourse, should be cut on the field side of the fence. These are usually made of a depth of about 3 ft. below the formation surface of the road, are about 1 ft. 6 in. in width at the bottom and about 4 ft. or 5 ft. wide at top, their actual size depending upon the nature of the surrounding country. "In crossing marshy land they should be made sufficiently deep and wide to obtain earth to raise the bed of the road, from side to side, 3 ft. higher than the natural surface, in order to compress the subsoil and *reduce its elasticity.*"

In addition to ordinary drainage, in marshy ground it will be necessary to excavate the surface material to a depth of 2 ft. or 3 ft., according to circumstances, and to fill in with clean sand or gravel, and so form an artificial base for the road. In some cases a sort of raft is formed of brushwood or dried peat, to carry the gravel or soil upon which the road metal is placed. The brushwood consists of long straight twigs, 10 ft. to 20 ft. in length, tied in bundles, called "*fascines,*" of from 9 in. to 12 in. in diameter. These are laid in alternate layers lengthwise and crosswise, forming a bed 18 in. in thickness, and are either fastened with pegs or the bound fascines are cut to permit of a compact bed being formed

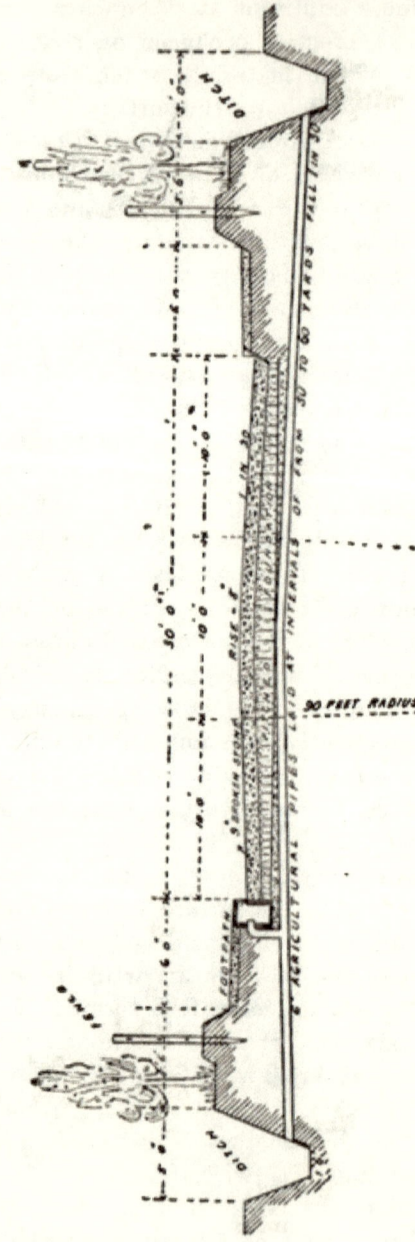

Fig. 14.—Cross-Section of a 30-ft. Country Main Road.

upon which the gravel is spread. The fascines, to prevent decay, should be permanently wet.

In cases where an open side drain cannot be adopted, as in passing through cuttings, near buildings, &c., a stone, brick or pipe covered drain must be made. These are frequently formed with a flat stone bottom and top, and stone sides built, as shown in Fig. 15, without mortar.

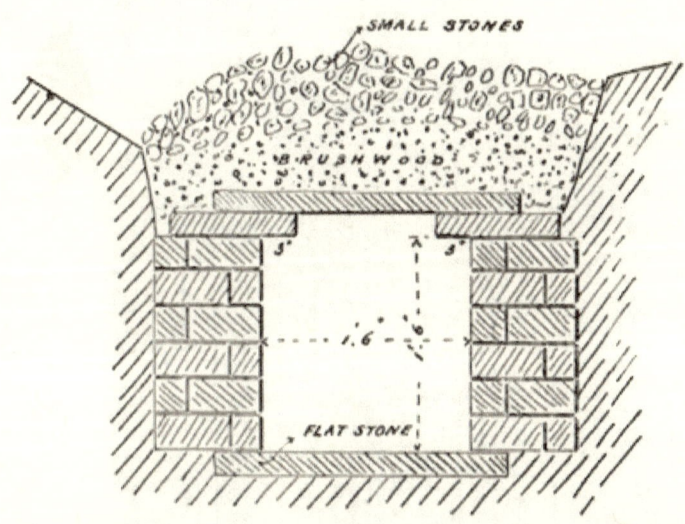

Fig. 15.—Stone Side Drain for Surface Drainage.

The water from the side channels of the roadway must be conveyed into the main side drains by small drains, frequently of agricultural pipes, formed at intervals of about 60 ft. underneath the footpaths. They should have an inclination to the side drains of about 1 in 30.

If the subsoil is a soft wet clay, or otherwise elastic nature, some additional precautions to ensure the stability of the foundation of the road will be necessary, such as the construction of *cross* or *mitre drains* from the centre of the road to the side main drain. When the "surface formation" has been prepared for the road materials trenches are cut across the road at an angle with its centre line at intervals, usually of about 60 yards, but much closer if the subsoil is very wet.

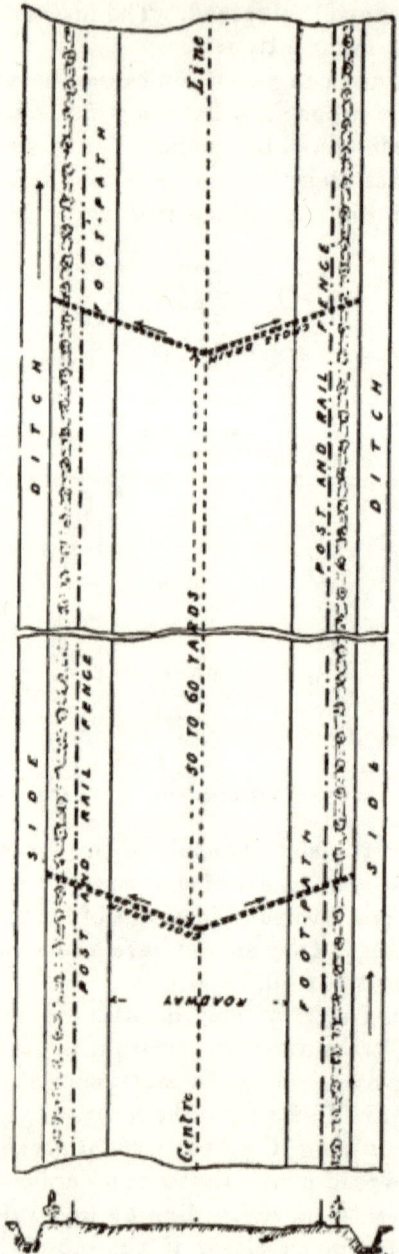

Fig. 16.—Plan of Road showing Cross Drains from Centre of Road to Side Main Drain.

The splay formed by these drains with the centre line will depend upon the inclination of the road. If it is level they may cross at right angles, but where there is an inclination they should form an angle of a flat V-shape, with the centre line as shown in Fig. 16. The trenches should be excavated to about 18 in. deep, 12 in. wide at the bottom and about 16 in. at the top. In the bottom an open channel, not less than 6 in. by 4 in., should be formed, as shown in Fig. 17, in

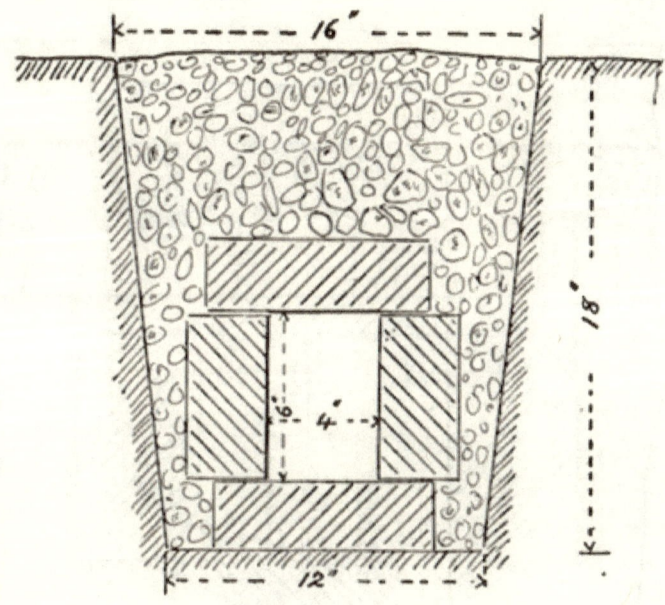

FIG. 17.—CROSS OR MITRE DRAIN.

brick or stone, with open joints and a covering of small stones filled in to the level of the *formation surface,* so as to be in direct contact with the road materials to draw off any water from them.

In lieu of the stone channel, shown in the figure, 4-in. or 6-in. agricultural pipes, with an inclination of about 1 in 30, may be used, the trench being filled in with small stones or hardcore as before.

Cross drains are also sometimes required to convey water

from one side of the road to the other. For instance, "when the road passes along the slope of a hill or mountain a great number of these drains are necessary to carry off the water that collects in the channel on *the side next the high ground*. They should be placed at from 50 yards to 100 yards distance from each other, according to the declivity of the hill, so that the *side channels* may not be cut by carrying water too far. In these situations *inlets* should be built of masonry, to carry the water from the side channel into the cross drains."*

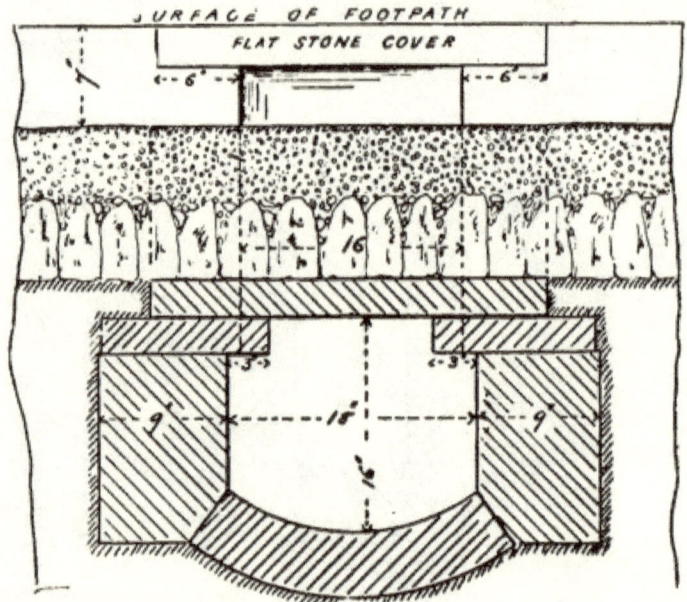

Section AB (*see* Fig. 19).

FIG. 18.—CROSS SECTION OF "INLET" TO CARRY WATER FROM SIDE CHANNEL TO CROSS DRAIN.

These inlets for country roads may be constructed in stone or brick, as shown in Figs. 18 and 19, and covered in with large flat stones, fixed with the upper face 6 in. above the level of the side channels. The mouth of the inlet may be provided with a cast-iron hinged grating, with bars ¾ in. wide by 2 in. deep and 1 in. apart.

---

* Sir H. Parnell.

Where inlets of the above description are formed the small agricultural pipe drains under the footpath from the side channel to the ditch, as previously referred to, may, of course, be then dispensed with.

Other arrangements for conveying away the drainage from roads are illustrated in Figs. 20 and 21. Methods similar to these were adopted on the Highgate Archway-road.*

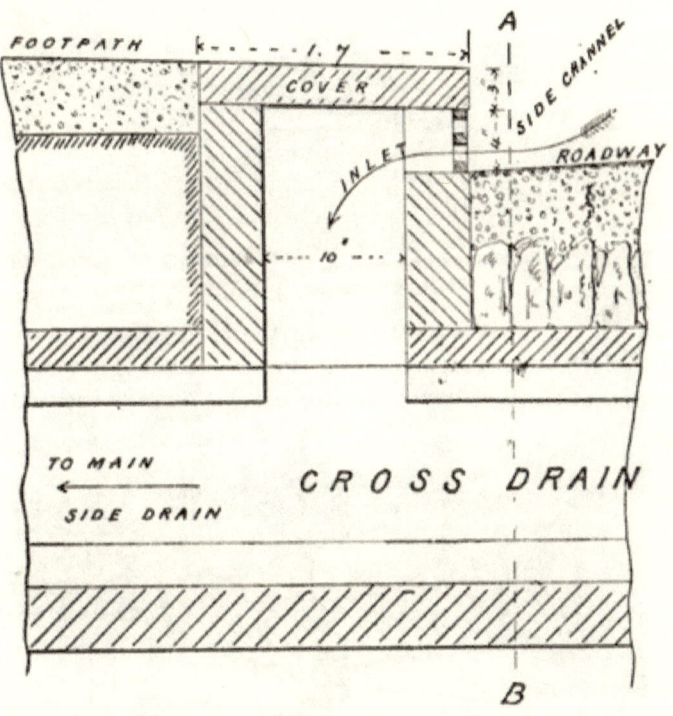

FIG. 19.—LONGITUDINAL SECTION OF CROSS DRAIN, SHOWING "INLET" FROM SIDE CHANNEL.

Road Drainage in Towns.—Town roads are drained into underground sewers, either specially constructed for surface water only or for the conveyance of sewage and surface water combined.

---

* See " A Treatise on the Principles and Practice of Levelling," by F. W. Simms, M.INST.C.E.

Fig. 22 illustrates a town road in cross section, showing a general arrangement of sewers and connections therewith for the removal of surface water.

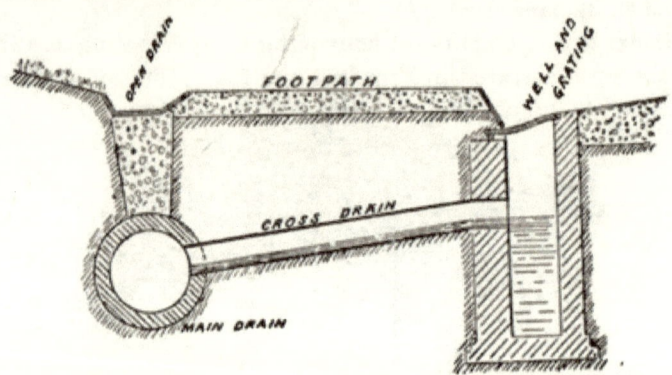

Fig. 20.—Another Mode of Removal of Surface Drainage from Roads.

Provision must here be made for the removal of a very much larger proportion of the rainfall than is necessary in the case of rural districts, and the dimensions and inclinations

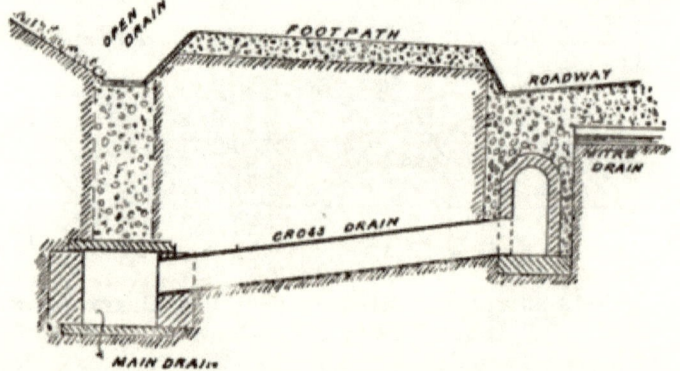

Fig. 21.—Removal of Surface Drainage from Roads, Alternative Arrangement.

of the sewers designed for its reception must, therefore, be suited to the heaviest rainfall likely to occur in a short space of time. In rural districts it is usual to estimate upon about

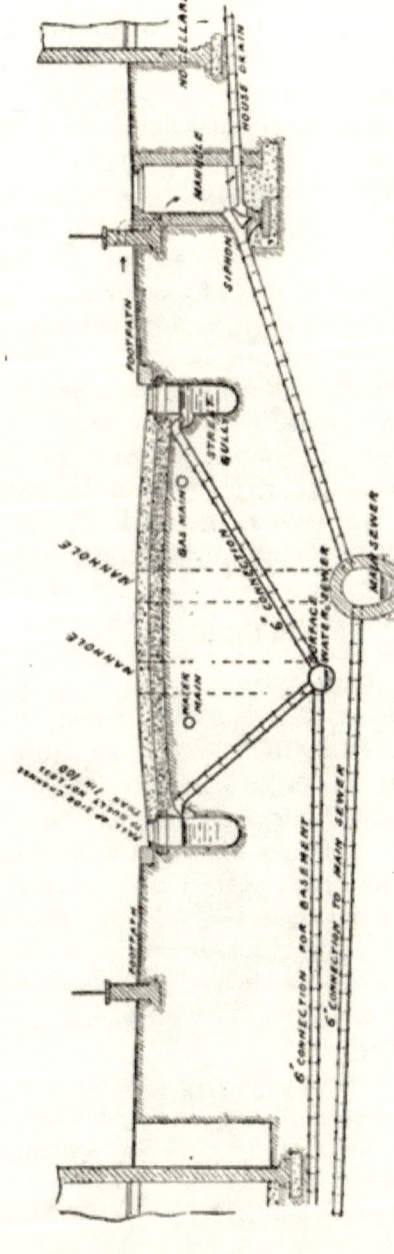

Fig. 22.—Cross Section of a Town Road, showing General Arrangement of Sewers, and Connections for Removal of Surface Water.

one-third of the rainfall directly entering the natural watercourses, about two-thirds entering the sewers in country towns, and as much as three-fourths in thickly-populated cities with well-paved streets, courts, yards, &c.

In London as much as 6 in. of rain have been known to fall in one and a half hours, although this, happily, is very exceptional; however, other observations in England as to the extent and frequency of storms have led to the statement of the following rule, that " when sewers are constructed to carry off storm waters they should be of a capacity to discharge a proportion of a 4-in. rainfall in twenty-four hours, varying according to the character of the district."

But upon this point there seems to be considerable difference of opinion. In connection with the main drainage of London Sir Joseph Bazalgette, in considering as to the amount of rainfall to be estimated upon as influencing the determination of the capacity of the sewers, "ascertained that there are about 155 days in the year upon which rain falls; of these there are about twenty-five days on which the rainfall amounts to $\frac{1}{4}$ in. in twenty-four hours, or at the average rate of $\frac{1}{100}$th part of an inch per hour." In a report (1858) of Messrs. Bidder, Hawksley and Bazalgette, the engineers appointed to inquire into this subject, it is concluded " that the quantity of rain which flowed off by the sewers was in all cases much less than the quantity which fell on the ground; and, although the variations of atmospheric phenomena are far too great to allow any philosophical proportions to be established between the rainfall and sewer flow, yet we feel warranted in concluding, as a rule of averages, that $\frac{1}{4}$ in. of rainfall will not contribute more than $\frac{1}{8}$ in. to the sewers; nor a fall of $\frac{4}{10}$ in. more than $\frac{1}{4}$ in. Indeed, we have recently observed rainfalls of very sensible amounts failing to contribute any distinguishable quantity to the sewers."

" To allow for exceptional cases of heavy and violent rainstorms, which have measured 1 in., and sometimes even 2 in., in an hour, overflow weirs, to act as safety valves, were constructed at the junction of the intercepting sewers with the main valley sewers, over which the *surplus* waters, largely diluted, flow through the original channels into the Thames. The sewers have been made capable of carrying off a volume equal to a rainfall of ¼ in. per day during the six hours* of maximum flow."†

The surface water falling upon the street (Fig. 22) is conveyed by means of the *side channels*, which have a fall of about 1 in 100, to *street gullies* fixed at intervals of from 50 yards to 60 yards apart, and which are connected with 6-in. salt-glazed stoneware socket pipes to the surface-water main sewer as shown.

Street gullies are variously formed, usually either of stone, brick, stoneware or iron. A large number of old gullies at present exist constructed in the form of square or rectangular pits, built in stone or brick, sometimes with a circular bottom, and oftentimes trapped either with a "dipstone" or by means of a stoneware siphon. Many have been put in without pretence either of being watertight or trapped, and have in this latter form been considered as efficient ventilators to the sewers. Modern views, however, on the question of sewer ventilation have, to a very large extent, abandoned so doubtful a practice as the use of the street gully as a ventilator, and the old untrapped brick or stone pits are giving place to a more sanitary arrangement.

The essential points for a good street gully are—

1. The area of open grating must be sufficient to cope with heavy and sudden rainfalls.
2. The grating must not admit of being easily choked on its surface by road *débris*, paper, straw, leaves, &c.
3. The gully must be of adequate capacity to retain sand, mud and other road detritus, and so prevent the entry of same into the sewers.
4. It must admit of being *easily cleared;* also of the con-

---

* The metropolitan sewers receive both sewage and surface water, and, as it is found in practice that a sewage flow is not uniform *throughout the twenty-four hours*, therefore provision was made, in determining the capacity of the sewers, for one half of the sewage to flow off within six hours of the twenty-four.

† "Drainage of Lands, Towns and Buildings," by G. D. Dempsey and D. K. Clark.

nection to the sewer being conveniently rodded in case of a stoppage. For these reasons the construction of the gully should be as simple as possible— several movable parts may be looked upon as a defect in design

5. There should be a deep water seal or trap to prevent the escape of sewer gas. Shallow water seals soon evaporate in dry weather.
6. The gully should afford no obstruction to the traffic, and the outlet should be so placed as to be free from danger of injury either by steam roller or heavy vehicles.
7. The gully must be of a material not easily damaged by scoops or other tools used in clearing it.

Doulton's improved street gully, or circular pot gully (Fig. 23), is a good form, of *simple* construction, and is very generally used. It is made of stoneware, in one piece, and is fixed as shown in the accompanying sketch with two courses of brickwork upon which rests the cast-iron grating. The gully is also made with a valve trap combined with the ordinary siphon, with a hole for inspection at the outlet.

Sykes' patent street gully is rectangular in form, with rounded angles, and is made of "granitic stoneware." The gully has a deep seal so arranged that when the road detritus is taken out it remains sealed. The outlet for connection to the sewer is kept low with a view to avoiding danger of fracture by steam rollers; it is also taken down in a vertical direction to avoid contact with gas and water mains, and a screw-plugged inspection eye is provided over the outlet for cleansing.

Crosta's patent surface-water gully is manufactured in cast iron of various capacities, from about $7\frac{1}{2}$ gallons up to 57 gallons. The gully has been largely adopted in London, Liverpool and other towns. It is provided with a double trap, whereby the escape of sewer gas is prevented even when the body of the gully is empty.

Wakefield's street gully is also a recent production, possessing a double water seal of from 2 ft. to 3 ft. according to the depth of the gully. It is made of stoneware, with capacities of from 45 to 60 gallons.

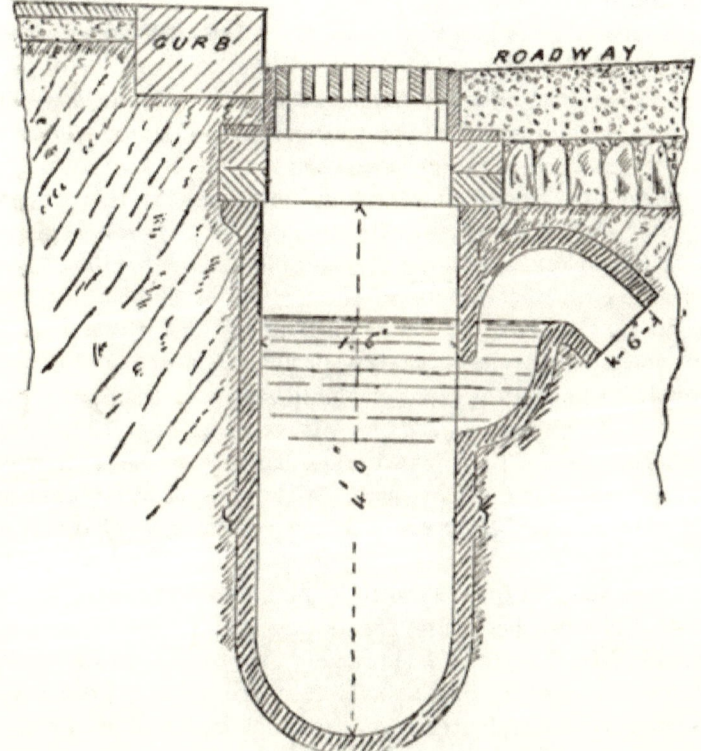

Fig. 23.—Stoneware Circular Street Gully, showing Fixing.

The Patent Victoria Stone Company manufacture a "catch-pit gully" with cascade trap of their stone material.

FENCING.

If a road is drained by means of an open ditch a fence will be necessary to prevent accidents between the road and the ditch; but, as far as the durability of the road itself is concerned, fences are best dispensed with. In all cases they should be as far from the sides of the road as possible, and should be kept low—the maximum height desirable, according to Telford, being 5 ft. Higher fences, and trees especially, seriously intercept the beneficial influence of the sun and wind, and therefore diminish their effect in producing evaporation.

Where stone is plentiful *walls* make the best fences. They may be built in dry rubble work without mortar.

An open *post-and-rail* fence, although very suitable as far as the durability of the road is concerned, is very liable to decay and damage, and therefore expensive to maintain.

Through country districts the most common fence is the *quickset hedge*, which, if properly planted and well attended to in its earlier stages, is very suitable and economical. The following specification as to the method of planting is given in Sir H. Parnell's "Treatise on Roads": "A ditch is to be cut and a bank raised, together occupying a space of 8 ft. in breadth, the ditch is to be on the field side of the bank, to be cut out of the natural ground, 4 ft. wide at top, 10 in. wide at bottom, and $2\frac{1}{2}$ ft. deep.* The bank is to be 4 ft. wide, and is to be raised by sods, with the green or swarded side out, to the height of 14 in. above the side channels of the road.

"Two rows of quicks are to be planted on the ditch side of the bank, a bed being first formed for them of good vegetable mould 15 in. deep and 18 in. wide. There are to be twelve plants set in every lineal yard; they are to have good roots, three years transplanted from the quick bed, and of a strong and healthy appearance.

"These quicksets are to be protected by two rows of posts and rails, three rails in each row; the posts to be of good oak, 5 ft. long, 5 in. deep by 3 in. wide, with large buts sunk 2 ft. in the ground.

"The rails are not to be more than 8 ft. long, to be $3\frac{1}{2}$ in. wide by $1\frac{1}{2}$ in. deep, of good elm, oak or ash timber.

"In each length of rails two centre posts at least 2 in. wide by $1\frac{1}{2}$ in. thick are to be driven into the ground and fastened to the rails with strong nails."

Quickset hedges should be clipped yearly in August or

---

* Where the soil is clay the drain should be 4 ft. deep.

September, and should be trimmed level and regular. The planting should be done during the spring or autumn.

"No matter whether a road lies high or low, if it is overhung by trees it will be found to be constantly damp; this is most noticeable in summer, when the trees are covered with foliage. A neatly-cut hedge, well set back from the travelling way of a road, is about the best form of fence that can be devised; but when overgrown they become not only an obstruction to the traffic, but they shut out the sun and wind from the road surface. Although provision is made by Act of Parliament* whereby owners who refuse to lop their trees or cut their hedges, on being notified, can be compelled to do so, yet summoning such persons to petty sessions is one of those unpleasant tasks which a county surveyor is anxious, if possible, to avoid."†

TRACTION ON ROADS.

The theoretically perfect road would require to possess the qualities of straightness, levelness, smoothness and hardness, or inelasticity in absolute perfection, and would require no tractive force to carry a load along it, for assuming the air resistance and the friction of the wheels on the axles to be *nil*, an impulse given at one end would convey the load to the other by its inertia alone. This principle of science, which was first laid down by Sir Isaac Newton, is known as the "first law of motion," and may be stated as follows: A body when once set in motion continues to move uniformly forward in a straight line by its inertia until it is compelled to change that state by the action of some external force.

The main external forces offering resistance to the motion of carriages upon roads are:—

Friction.—"The resistance which friction occasions partakes of the nature of the resistance of fluids; it consists of the consumption of the moving force, or of the horses' labour,

---

* Highway Act, 1835 (5 and 6 Will. IV., chap. 50, ss. 65, 66), and sec. 48 and 49 Vict., chap. 13.

† "Proceedings of the Association of Municipal and County Engineers," vol. xviii., paper on "Road Maintenance," by R. H. Dorman.

occasioned by the soft surface of the road and *the continual depressing of the spongy and elastic sub-strata of the road.** There is also a resistance arising from the *friction* of the axletrees of the vehicles.

Collision, as occasioned by hard substances, stones or other inequalities of the road surface.

Gravity, when a road is not horizontal, offers considerable resis'ance, proportionate to the steepness of the incline. Its effect, however, when the road is level is immaterial, as it acts in a direction perpendicular to the plane of the horizon, and neither accelerates nor retards the motion.

Air.—The resistance arising from the force of the wind will vary with the velocity of the wind, with the velocity of the vehicle, with the area of the surface acted upon, and also with the angle of incidence of direction of the wind with the plane of the surface.

The following table† gives the force per square foot for various velocities :—

Velocity of wind in miles per hour.	Force in lb. per square foot.	Description.
15	1·107	Pleasant breeze
20	1·968	Brisk gale
25	3·075	
30	4·428	High wind
35	6·027	
40	7·872	Very high wind
45	9·963	
50	12·300	Storm

To find the force of the wind†—

Assuming

V = velocity of wind in feet per second,
v = velocity in miles per hour,
P = pressure in lb. per square foot,
x = angle of incidence of direction of the wind with the plane of the surface when it is oblique,

---

* Prof. Leslie's " Elements of Natural Philosophy."
† Molesworth's " Engineers' Pocket-Book."

Then
$$P = \cdot 002288 \ V^2$$
$$P = \cdot 00492 \ v^2,$$
$$P = \cdot 0023 \ V^2 \times \text{sine } x.$$

The steepest inclination prevailing generally throughout a line of roadway is known as the "*ruling gradient*," and should be exceeded only under exceptional circumstances, such, for example, as where additional motive power or lighter loads are obtainable. In fixing this gradient the amount of motive power available in ascending and the avoidance of waste of power in descending are two main considerations. Also, there should be no waste of mechanical energy "through the necessity of using brakes, or of backing the prime mover, in order to prevent excessive acceleration of speed in descending the ruling gradient . . . . When the traffic is heavier in one direction than in another the ruling gradient in the direction of the ascent of the lighter traffic may be the steeper . . . . The less the proportion of the resistance on a level to the load the flatter must be the ruling gradient, and the flatter the ruling gradient is the heavier are the works, and the more difficult is it to lay out the line. Such, for example, is the case with railways as compared with roads."

"Telford estimated the average resistance of carriages on a level part of a good broken stone road at one-thirtieth of the gross load"; and, according to the following principle— viz., that *the sine of the angle of inclination of the ascent should (if possible) not exceed the proportion of the resistance to the load on a level*—" he assigned 1 in 30 as the ruling gradient which ought, as far as possible, to be adopted on a turnpike road.

"If the tractive force which a horse can exert steadily and continuously at a walk be estimated at 120 lb., the adoption of a ruling gradient of 1 in 30, the resistance on a level being one-thirtieth of the load, ensures that each horse shall be able to draw up the steepest declivity of the road a gross load of
$$\frac{120 \times 30}{2} = 1,800 \text{ lb.}$$

"A horse can exert for a short time an effort two or three

times greater than that which he can keep up steadily during his day's work, and thus steeper ascents for short distances may be surmounted.

"In the roads laid out by Telford the ruling gradient of 1 in 30 is adhered to wherever it is practicable to do so; and sometimes considerable circuits are made for that purpose. Occasionally, however, he found it necessary to introduce steeper gradients for a short distance, such as 1 in 20 or 1 in 15."*

The following observations by Mr. Thomas Codrington † bearing upon this point will also be of interest :—

"On a level macadamised road in ordinary repair the force which the horse has to put forth to draw a load may be taken as one-thirtieth of the load. But in going uphill the horse has also to *lift* the load, and the *additional* force to be put forth on this account is very nearly equal to the load drawn divided by the rate of gradient. Thus, on a gradient of 1 in 30 the force spent *in lifting* is one-thirtieth of the load, and in ascending a horse has to exert *twice* the force required to draw the load on a level. In descending, on the other hand, on such a gradient the vehicle, when once started, would just move of itself without pressing on the horse. A horse can without difficulty exert twice his usual force for a time, and can therefore ascend gradients of 1 in 30 on a macadamised surface without sensible diminution of speed and can trot freely down them. These considerations have led to 1 in 30 being generally considered as the ruling gradient to be aimed at on first-class roads."

Another authority, M. Dumas, recommended‡ 1 in 50 as the maximum rate of inclination, stating, as the result of experience, that "on broken stone roads in perfect condition the resistance to traction is one-fiftieth of the gross load, for which the angle of repose is 1 in 50"; also "that for the ascent of an incline of 1 in 50 the traction force required is just double that which is required on the level."

---

* *See* Prof. Rankine's "Manual of Civil Engineering" (19th edition), from which the foregoing remarks are abstracted.
† "*Encyclopædia Britannica.*"
‡ "*Annales des Ponts et Chaussées*," 1843.

"Sir John Macneill in 1836 maintained that no road was perfect unless its gradients were equal to or less than 1 in 40. In thus limiting the *ruling gradient* to 1 in 40 he justifies the assertion by the much greater outlay for repair on roads of steeper gradients. For instance, he adduces as a fact not generally known that if a road has no greater inclinations than 1 in 40 there is 20 per cent. less cost for maintenance than for a road having an inclination of 1 in 20. The additional cost is due not only to the greater injury by the action of horses' feet on the steeper incline, but also to the greater wear of the road by the more frequent necessity for sledging or braking the wheels of vehicles in descending the steeper portions."*

The following empirical formulæ† for calculating the resistance to traction on *level* roads have been given by Sir John Macneil for a common *stage waggon* (goods traffic):—

$$P = \frac{W+w}{93} + \frac{w}{40} + cv. \qquad (1)$$

For a *stage coach* (passenger traffic)—

$$P = \frac{W+w}{100} + \frac{w}{40} + cv. \qquad (2)$$

where
P = the force in lb. required to move the carriage along a horizontal road of given description at the given velocity,
W = the weight of the carriage or tare in lb.,
w = the weight of the load in lb.,
v = the velocity in feet per second,
c = a constant number depending upon the nature of the surface over which the carriage is drawn.

The value of c for the following kinds of road surfaces is:—

On a timber surface ... ... $c = 2$
On a paved road ... ... $c = 2$
On a well-made broken stone road in a dry, clean state ... ... $c = 5$

---

* Law and Clark on "Roads and Streets."
† Sir Henry Parnell, "Treatise on Roads."

On a well-made broken stone road
covered with dust ... ... $c = 8$
On a well-made broken stone road wet
and muddy ... ... $c = 10$
On a gravel or flint road in a dry, clean
state ... ... $c = 13$
On a gravel or flint road in a wet and
muddy state ... ... $c = 32$

In going up an *incline* a horse, as already stated, has to exert an amount of *additional force* in lifting the load, which must be added* to the resistance on the level, and the above formulæ are therefore modified accordingly.

The accompanying diagram (Fig. 24) illustrates the case of a vehicle sustained in equilibrium upon an inclined plane by

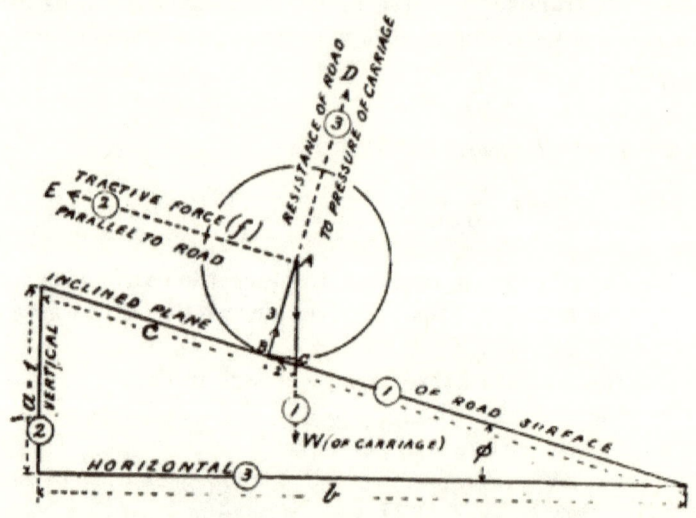

Fig. 24.

the three forces acting at point A in directions as indicated—viz.:—

Force AW representing the weight of the carriage and load, or the force of gravity acting vertically downwards.

---

* Or subtracted if descending a hill.

Force A E sustaining the carriage upon the incline and acting parallel to the road surface.

Force A D, the resistance of the road surface to the pressure of the carriage, acting at right angles to the inclined plane.

These forces, being in equilibrium, are proportionately represented by the three sides of the triangle ABC, which are parallel respectively to the forces named; and, as the angle ABC = angle formed by a and b (both being right angles), and the angle ACB = angle formed by a and c (Euclid I., 29), ∴ angle CAB = angle c and b, and the triangles A B C and $a\ b\ c$ are "similar," the sides of the latter being, therefore, also proportional, each to each as correspondingly numbered in the diagram, to the three forces in question, from which it is clear that—

W (weight of loaded carriage) : c :: Force AE : a

Therefore (denoting the force AE by f, and b being a horizontal length of road of which the rise $a = 1$)—

$$\frac{c \times f}{a} = W,$$

that is—

$$f = W \times \frac{a^*}{c} = W \cdot \sin. \phi \qquad (3)$$

Also, W : c :: Pressure or resistance AD : b

Therefore (denoting pressure AD by p)—

$$\frac{c \times p}{b} = W,$$

that is—

$$p = W \times \frac{b\dagger}{c} = W \cdot \cos. \phi \qquad (4)$$

NOTE.—p, it will be seen from formula (4), is a little less than the actual weight of the loaded carriage, but, except on steep inclinations, may be neglected in the following formulæ and taken as being equal to the weight.

---

* $\dfrac{a}{c} = \dfrac{\text{height}}{\text{length}}$ or $\dfrac{\text{perpendicular}}{\text{hypotenuse}} = \sin. \phi.$
  ϕ = angle formed by cb (see figure).

† $\dfrac{b}{c} = \dfrac{\text{base}}{\text{hypotenuse}} = \cos. \phi.$

The resistance to traction in *ascending* or *descending* an inclined plane will therefore be P $\pm$ f*, and introducing the value of f (Formula 3) into formula (1), and allowing for the reduction of pressure p (Formula 4) due to the incline, the formula for resistance now becomes—

For a common stage *waggon* ascending or descending an incline—

$$P = \left(\frac{W+w}{93} + \frac{w}{40}\right) \cdot \cos \phi \pm \left\{(W+w) \cdot \sin \phi\right\} + cv \quad (5)$$

For a stage *coach*—

$$P = \left(\frac{W+w}{100} + \frac{w}{40}\right) \cdot \cos \phi \pm \left\{(W+w) \cdot \sin \phi\right\} + cv \quad (6)$$

EXAMPLE.—Calculate the force required to move a stage coach weighing 1 ton, and having a load of 2 tons, at a velocity of 6 miles per hour along a broken stone road with its surface wet and muddy, and which has an inclination of 1 in 40.

Gross load = 2240 lb. + 4480 lb. = 6720 lb.

Velocity of 6 miles per hour = $\frac{528}{60}$ = 8·8 ft. per second.

Value of c for broken stone road, wet and muddy = 10.

For inclination of 1 in 40 the angle ϕ (*see* Fig. 25) is 1° 25′ 57″, the nat. sin. of which is ·0247+; and the nat. cos. ·9996+.

Therefore, when *ascending* the incline—

$$P = \left(\frac{2240+4480}{100} + \frac{4480}{40}\right) \cdot 9996 + \left\{(2240 + 4480) \cdot 0247\right\}$$
$$+ (10 \times 8·8) = 179·13 + 165·98 + 88 = \underline{\underline{433·11 \text{ lb.}}}$$

When *descending*—

$$179·13 - 165·98 + 88 = \underline{\underline{101·15 \text{ lb.}}}$$

Mr. D. K. Clark, M.I.C.E., in an analysis of the "Rolling or Circumferential Resistance of Wheels,"† states it as being "equal to the product of the load by the third of the semi-

---

* + f when the load is being drawn *up the hill*, and − f when *descending* same.

† In the "Construction of Roads and Streets," by Law and Clark.

chord (of the submerged arc of the wheel), divided by the radius of the wheel.

That is—

$$\text{Tractive force} = \tfrac{1}{3}\, \frac{W \times \text{semichord of submerged arc of wheel}}{\text{radius of wheel}}.$$

"This question is," he adds, "no doubt applicable, with a sufficient degree of accuracy, for any real needs for calculating the resistance of gravel, loose stones, soft earth or clay."

Also, he deduces the following rule: That "the circumferential or rolling resistance of wheels to traction on a level road is inversely proportional to the cube root of the diameter;" so that "to reduce the rolling resistance of a wheel to one-half, for instance, the diameter must be enlarged to eight times the primary diameter." M. Morin, however, deduces from his experiments upon resistance to traction that "the resistance varies simply in the inverse ratio of the diameter of the wheel."

M. Dupuit, as a result of experiments carried out by him, deduced that on macadamised roads in good condition and on uniform surfaces generally—

1. "The resistance to traction is directly proportional to the pressure.
2. "It is independent of the width of the tyre.
3. "It is inversely as the square root of the diameter.
4. "It is independent of the speed."

"M. Dupuit admits that on paved roads, which give rise to constant concussion, the resistance increases with the speed, whilst it is diminished by an enlargement of the tyre up to a certain limit."*

There is also an occasional resistance produced by *collision* with loose stones or other hard substances, which give a sudden check to the horses, depending upon the height of the obstacle—the momentum destroyed being oftentimes very considerable.

The pressure which is thus brought to bear upon a stone or other obstacle in being surmounted is equal to the joint

---

* The "Construction of Roads and Streets," by Law and Clark.

action of the tractive *power* and *weight* of the loaded carriage.

"The injury which a road sustains by this pressure acting on a small point (upon the obstacle) and in an oblique direction is very great; but it is not alone in this that the road suffers, the force with which the wheel strikes the surface in its descent from the top of the stone is considerable and would soon wear a hole in the hardest road. But it must be observed that a carriage mounted on proper *springs* will be drawn over an obstacle of this kind with much less power than if the carriage had no springs, for the springs allow the wheels to mount over the obstacle *without raising the body of the carriage* and its load with it to the same height. Upon this principle alone it is that carriages mounted on proper springs are easier moved than those without springs, and, for the same reason, springs are more necessary on rough and uneven roads than on smooth ones, and in proportion to the roughness of the roads should the springs be free and elastic."

"By enlarging the diameter of the wheel the power required to draw it over an obstacle would be diminished, and should the weight of the wheel remain the same the power will decrease nearly as the diameter of the wheel increases."

"If the power be applied to the axle in a direction not parallel to the horizon, but inclining upwards, the resistance will be diminished, or a less power will be required, for the leverage by which the power acts is increased, while the leverage by which gravity acts is decreased, until the line of draught forms a right angle with the line drawn from the centre of the axle to the point of contact, in which case the power is a minimum. On the contrary, if the direction of the line of draught be inclined downwards from the horizontal, the leverage by which it acts will be diminished, and consequently the power must be increased as the direction of the line of draught falls below the horizontal."*

The general deductions to be drawn from the experiments

---

* Sir Henry Parnell's "Treatise on Roads."

made by M. Morin upon the resistance to the traction of vehicles on common roads are :—

1. "The resistance to traction is directly proportional to the load, and inversely proportional to the diameter of the wheel (compare with deduction by Mr. D. K. Clark above).
2. "Upon a paved or a hard macadamised road the resistance is independent of the width of the tyre when this quantity exceeds from 3 in. to 4 in.
3. "At a walking pace the resistance to traction is the same, under the same circumstances, for carriages with springs and for carriages without springs.
4. "Upon hard macadamised roads and upon paved roads the resistance to traction increases with the velocity, the increments of traction being directly proportional to the increments of velocity above the velocity 3·28 ft. per second, or about $2\frac{1}{4}$ miles per hour. The equal increments of traction thus due to equal increments of velocity are less, as the road is smoother and as the carriage is less rigid or better hung.
5. "Upon soft roads of earth or sand, or turf, or roads fresh and thickly gravelled, the resistance to traction is independent of the velocity.
6. "Upon a well made and compact pavement of hewn stones the resistance to traction at a walking pace is not more than three-fourths of the resistance upon the best macadamised roads under similar circumstances. At a trotting pace the resistances are equal.
7. "The destruction of the road is in all cases greater as the diameters of the wheels are less, and it is greater in carriages without than with springs."*

Sir John Macneill, in connection with his investigations as to the best plan to be adopted for improving that portion of the Holyhead-road through the Stowe Hill Valley, in order to ascertain what sum of money should be laid out on the works so as to produce the most advantageous result, and to decide

---

* "The Construction of Roads and Streets," by Law and Clark.

whether or not the same sum of money would produce an equally beneficial improvement if laid out in raising the valley without lowering the summits, or in lowering the summits without raising the valley, states * that, "to arrive at an accurate result in this and similar investigations it is necessary to know correctly the expense of horse labour under the varying circumstances of velocity and force of traction on different inclined planes, and also the draught of carriages and the ratio of the increase of draught in consequence of increase of velocity."

These circumstances he ascertained from experiments made on the Holyhead road, from which formulæ were deduced, and a system of tables calculated therefrom, showing the expense of drawing a given weight with a given velocity over every rate of acclivity and declivity and length of inclined plane.

He also further states that "by means of these tables the expense of drawing a ton weight over any line of road may be determined with great accuracy. Hence, all that is necessary in the present investigation is to calculate by the tables the expense of transporting a ton weight over the existing line of road and also over the proposed improvements. The difference will be the saving in expense of drawing a ton with the given velocity over the proposed improvement. This, multiplied by the number of tons that pass over the road each day and by the number of days in the year, will give the annual saving which, compared with the interest of the money necessary to be expended in making the improvement, will clearly show whether the saving in horse labour is commensurate with the proposed expense. By applying the same criterion to each of the proposed plans it will at once be made evident which of them should be adopted, as that which would produce the most beneficial results at the smallest expense."

As an example of the application of the above method the following figures may be quoted :—

---

* Sir Henry Parnell's "Treatise on Roads."

From the tables the mean expense of drawing a ton over an existing line of road between two given points, in *both* directions, is 82·0647d., but over the proposed improved line it is 76·1724d., thus showing a saving in horse labour of 5·8923d. per ton; and for a daily traffic of 170 tons the saving will be £4 3s. 6d. per diem, which at 5 per cent. is interest for £30,310 10s. The estimate for making the proposed improvement was £23,757. "The difference between the amount of the estimate and the saving to the public by the proposed improvement is therefore £6,553, which is the actual sum the public would gain by this improvement, supposing the present traffic to continue; if the traffic increased the saving would be still more."

Draught.—The draught of horses upon different road surfaces will be affected by the amount of resistance to traction offered by the road surface, by the rate of inclination and amount of foothold afforded, also by the speed, diameter of the wheels and their friction upon the axles, and by the force of the wind.

The following* are the results of experiments on London street surfaces made by Sir J. N. Bazalgette with Easton & Anderson's horse dynamometer—the gross load was 4 tons and the speed 2 to 6 miles per hour:—

	Tractive force on the level.			
Macadam surface	40·7	to	44·29	lb. per ton.
Asphalte†	39·0	„	39·32	„ „
Wood	33·62	„	36·63	„ „
Granite setts	26·2	„	27·0	„ „

The following comparative statement‡ shows the *traction upon level roads* formed of different surfaces as described. *Asphalte* is taken as the standard of excellence = 1:—

---

* From "*Encyclopædia Britannica*," Art. "Roads and Streets," by Mr. T. Codrington.

† It is very surprising to find asphalte so high.

‡ "Municipal and Sanitary Engineers' Hand-Book," by Mr. H. P. Boulnois.

Asphalted roadway	...	...	...	...	1·0 —
Paved roadway, dry and in good order...				...	1·5 to 2·0
,, ,, in fair order	...	...	...		2·0 ,, 2·5
,, ,, but covered with mud...				...	2·0 ,, 2·7
Macadamised roadway, dry and in good order...					2·5 ,, 3·0
,, ,, ,, in a wet state	...			...	3·3 —
,, ,, ,, in fair order	...			...	4·5 —
,, ,, ,, but covered with mud...					5·5 —
,, ,, ,, with the stones loose				...	5·0 ,, 8·2

WIDTH, CROSS-SECTION AND GRADIENT.

Width.—Roads should be made of adequate width, having regard to the locality, the amount of traffic, nearness or otherwise to large towns, and the cost of construction and maintenance. The first cost of a wide road, as compared with that of a narrower one, will be approximately proportionate to its width; the cost of maintenance, however, will depend not only upon the superficial area, but also very largely upon the nature and amount of the traffic accommodated; so that, whether the road be wide or narrow, the quantity of material used will not vary much—the labour only being increased in the case of a wide road.

To form a correct opinion as to the wearing effect of traffic upon the roadway surface, traffic has been reduced by Mr. Deacon to a standard expressed in terms of *tons weight of traffic per yard of width of roadway per annum.**

The width of the carriageway of ordinary English turnpike or main roads is about 30 ft., which should be the minimum width for all roads between large towns; also, an additional 6 ft. is required for a footpath on one side. On approaching large cities this breadth must oftentimes be increased to 40 ft. or 50 ft., with footpaths 7 ft. or 8 ft. wide on each side; but no general rule can be laid down.

The comparative widths of various classes of roads will be seen from the following statement:—

---

* Paper by Mr. Deacon, "Proceedings of the Institution of Civil Engineers," vol. lviii.

Description.	Width of Roadway.	Width of Footpaths.	Total Width.
Country land	12 ft. to 15 ft.	None	—
Ordinary country road	15 ft. to 20 ft.	None	—
Country road between large towns	30 ft.	6 ft. (one side only)	36 ft.
Near large towns	45 ft.	6 ft. (both sides)	57 ft.
Common city road	40 ft.	10 ft. (both sides)	60 ft.
Suburban road	25 ft.	7 ft. 6 in. (both sides)	40 ft.
Chelsea, London	26 ft.	7 ft. (both sides)	40 ft.
King William-street, London	33 ft.	8 ft. 9 in. (both sides)	50 ft. 6 in.
Southwark-street, London	46 ft.	12 ft. (both sides)	70 ft.
Hill roads in India	20 ft.		
Roads in Central India	First class, 30 ft.; second class, 24 ft.; third class, 20 ft.		
Grand Trunk Road, Bengal	Standard width, 40 ft.; with the central 16 ft. metalled with broken granite.		
Belgium, type-section	29 ft. 6 in., with an additional 5 ft. on each side for open ditch. The central 9 ft. 10 in. of roadway of stone pavement on concrete.		
French, " "	32 ft. 10 in., with an additional 5 ft. on each side for open ditch. The central 19 ft. 8 in. of roadway metalled with broken stone.		
Rue de Rivoli, Paris	46 ft.	16 ft. 5 in. (both sides)	78 ft. 10 in.

In England the requirements of different towns as to the width of streets varies considerably. In Birmingham the minimum is 50 ft., whilst in Manchester, Liverpool and Bristol it is only 30 ft.

The Parliamentary regulations prescribe the following widths, heights, &c.,* for those portions of public roads in Great Britain interfered with by the crossing of railways:—

Description.	Turnpike Road.	Public Road.	Occupation Road.
	Feet.	Feet.	Feet.
Clear width of under bridge or approach ...	35	25	12
Clear height of under bridge for a width of 12 ft...	16	—	—
Ditto for a width of 10 ft....	—	15	—
Ditto for a width of 9 ft. ...	—	—	14
Ditto at springing ...	12	12	—
Over bridge: Height of parapets	4	4	4
Approaches: Inclination ...	1 in 30	1 in 20	1 in 16
Ditto: Height of fencing	3	3	3

The Public Health Act, 1875, sec. 144, vests urban authorities in the powers of surveyors of highways under the Highway Act, 1835,† and sec. 80 of which latter enactment provides that "the surveyor shall maintain every *public cartway* leading to any market town *20 ft.* wide at least, every *public horseway 8 ft.* wide at least, and every *public footway* by the side of any carriageway or cartway *3 ft.* wide at least, if the ground between the fences will admit thereof."

At the present time in urban districts the actual details of the construction of roadways are usually regulated by by-laws made by the highway authority of the district, under sec. 157 of the Public Health Act, 1875,‡ which provides that "every urban authority may make by-laws with respect to the *level, width* and *construction* of new streets."

---

*Molesworth's "Engineers' Pocket-Book." *See* also " Standing Orders of Lords and Commons" (Waterlow & Sons) and Hodges' "Treatise on the Law of Railways" (Sweet & Maxwell).

† 5 and 6 Will. IV., c. 50.

‡ Or, in some cases, it may be under a local Act.

CROSS-SECTION AND GRADIENT.

Cross-section.—The transverse section of a good country road generally comprises a carriageway, a footpath on one or both sides, fences and side ditches (*see* Fig. 14); the section being, of course, necessarily modified when the road is in embankment or cutting.

The best form of cross-section or contour to be given to a roadway is a subject upon which there is considerable diversity of opinion, but, as pointed out* by Mr. Boulnois (city engineer, Liverpool), "for all practical purposes evenness of surface and regularity of section in a macadamised roadway are of more importance than the slight difference between straight lines and curves, which might only tend to confuse the workmen." It is, however, of great importance, as regards the durability of the road, that the form of cross-section shall be such as will ensure all water falling upon the roadway surface being immediately drained away to the side channels, and for which purpose only the surface has to be made convex. The evils, as formerly indulged in, of making roads with an excessive convexity—"hog-backed" or "barrelled" as it is termed—have already been pointed out.

Mr. Macadam considered that "a road should be as flat as possible with regard to allowing the water to run off it at all, because a carriage ought to *stand upright in travelling* as much as possible. He also says,† "I have generally made roads 3 in. higher in the centre than I have at the sides when they are 18 ft. wide; if the road be smooth and well made, the water will run off very easily in such a slope.

Telford recommended that the convexity of the roadway should be of the form of a very flat ellipse; he prepared the surface formation nearly level transversely and laid in a pitched foundation or "bottoming," 7 in. deep in the centre of the roadway and 4 in. deep at the sides, thus obtaining part of the convexity of the finished surface.

Mr. Walker preferred two straight lines inclined about 1 in 24 towards the sides for the haunches of the roadway and

* "The Municipal and Sanitary Engineers' Hand-Book."
† Parliamentary Report on the Highways of the Kingdom, 1819.

joined at the crown by a short curve, but the objection to this is that the flat sides very soon wear hollow. In respect to the form of cross-section, Mr. Walker stated* that "a road much rounded is dangerous, particularly if the cross-section approaches towards the segment of a circle, the slope in that case not being uniform, but increasing rapidly from the nature of the curve as we depart from the middle or vertical line. The over-rounding of roads is also injurious to them, by either confining the heavy carriages to one track in the crown of the road or, if they go upon the sides, by the greater wear they produce from their constant tendency to move down the inclined plane, owing to the angle which the surface of the road and the line of gravity of the load form with each other; and, as this tendency is perpendicular to the line of draught, the labour of the horse and the wear of the carriage wheels are both much increased by it." To facilitate drainage Mr. Walker also recommended that every part of a road, wherever possible, should have a longitudinal declivity of at least 1 in 80; and "when this cannot be obtained, owing to the extreme flatness of the country, an artificial inclination may generally be made. When a road is so formed every wheel-track that is made, being in the line of inclination, becomes a channel for carrying off the water much more effectually than can be done by a curvature in the cross-section or rise in the middle of the road, without the danger or other disadvantages which necessarily attend the rounding a road much in the middle."‡

A cross-section of the form of a *very flat ellipse* is, generally speaking, undoubtedly the best, and has the advantage over that consisting of a segment of a circle of giving somewhat flatter side slopes, without, at the same time, having the objection of liability to wearing hollow as when the haunches are formed by straight lines as advised by Mr. Walker.

A road laid practically level obviously requires a greater convexity of cross-section than when it has a longitudinal inclination; also, a new road should be more convex than

---

* Parliamentary Report, 1819.

G²

the intended finished contour, as by rolling and traffic it will soon spread towards the sides and flatten at the crown.

Mr. T. Codrington remarks* that the arc of a circle is often used and is good, but a curve more convex at the centre than towards the sides is best. The "*rise*" in the curve from the sides to the centre need not exceed one-fortieth of the width, and one-sixtieth is generally enough on well-kept roads; and if seven-eighths of the total rise are given at one-fourth of the distance from the centre to the sides, and five-eights at one-half the distance, a curve of suitable form will be obtained. It is best to obtain the convexity by rounding the "formation surface" and giving a uniform thickness of coating. When there is no curb there should be a "*shouldering*" of sods and earth on each side to keep the road materials in place and to form with the finished surface the water tables or side channels in which the surface drainage is collected, to be conveyed by outlets at frequent intervals to the side ditches.

The whole width of a roadway should be formed uniformly both as regards quality of materials and thickness of coating. The practice of making the centre 12 ft. or so of some hard material, such as granite, trap-rock or whinstone, and the side slopes of a weaker substance, as gravel, is a bad one, as the sides soon get out of repair and are quite unfit for the transit of heavy traffic.

For a suburban macadamised roadway the accompanying detailed transverse section (Fig. 25) illustrates a good form which may be adopted with advantage in residential areas where the traffic is not heavy. As regards the dimensions and details of construction the figure speaks for itself; the formation surface is parallel with the finished surface of the roadway, the "crown" of which is level with the heel of the footpath. If the ground is naturally wet, or contains springs, 3-in. or 4-in. agricultural drain pipes may be laid under the "side channel," as shown in the figure, and discharged into the street gullies.

---

* "Encyclopædia Britannica," article "Roads and Streets."

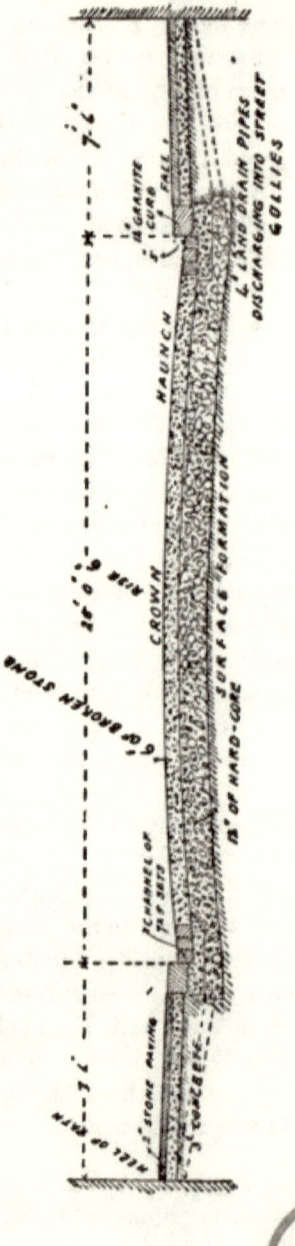

Fig. 25.—Cross-Section of Suburban Macadamised Roadway.

The experience of county surveyors as regards the cross-section of modern macadamised roads cannot fail to be of interest. Mr. R. H. Dorman, county surveyor, Armagh, Ireland, says,* "the cross-section of a road should vary with its hardness and with its longitudinal slope. On level roads 1 in 24 is a suitable transverse slop, where *limestone* or *gravel* are the materials used, 1 in 30 for *slatestone* and 1 in 36 for hard *whinstone* roads. On hills the camber should be increased—for instance, on an incline of 1 in 30 with transverse slope of 1 in 30, if the surface is perfectly smooth, the water will flow off the road into the side channels at an angle of 45°; but as all roads tend to become tracked longitudinally, the transverse slope on hills should be made greater than the longitudinal slope. For the cross-section of a road *a flat elliptical curve* for three-fourths of the width is preferable and the shoulders well filled out."

The above experience is supported on many points by that of Mr. E. B. Ellice-Clark, M.I.C.E., county surveyor for Sussex (west), who remarks that† "in some counties roads are 'hogbacked' to such an extent that vehicles cannot pass over them with safety except in the centre; the converse of this is found in leaving a road with a level transverse section so much out of repair that it becomes concave instead of being convex. It is obvious that if the versed sine of the section is too great vehicles will follow the same track—*viz.*, the crown of the road, which soon becomes worn out, while, if water is allowed to remain in the road, it will be destroyed by the weather. The proper fall for roads repaired with *flints* or similar material is 1 in 24 from the centre, with broken *granite* 1 in 36. The intermediate gradients between these two are those suitable for materials that are harder than flints and softer than granite, as for instance, hard mountain limestone, which will take its place at about 1 in 30. 'Barrelling' a road is a mere excuse for not keeping the surface in proper repair."

---

* "Proceedings of the Association of Municipal and County Engineers," vol. xviii.

† "Proceedings of the Association of Municipal and County Engineers," vol. xii.

In setting out and forming the cross-section and also in repairing roads it was formerly the practice to use a complicated "plummet rule" or "working level," which consisted of a long horizontal bar (equal in length to half the width of the road), fitted with a plummet and four gauges moving perpendicularly in dove-tailed grooves cut in the horizontal bar and fixed with thumbscrews for adjustment as required. These are now dispensed with, and three "*boning-rods*," together with a long straight edge and a spirit level, are found sufficient, with a good eye, to obtain a sightly section, both transversely and longitudinally.

Gradient.— Certain general observations as to "*ruling gradients*" and inclination *versus* length, curves, &c., have already been made when considering the question of *route* for a proposed line of communication.

The maximum inclination which should be given to a roadway ought not, if possible, to exceed the limiting angle of resistance for the materials forming its surface, the safe maximum gradient depending, therefore, upon the nature of the road surface and also upon the force required to propel a given load upon it.

By "*limiting angle of resistance*" is here meant that angle or inclination of a road at which a carriage, when once set in motion upon it, would just continue to roll down the incline by the action of the force of gravity alone.

Should the inclination exceed this the carriage will move at an accelerated rate, and, unless checked by the use of a drag* or some other means, the horses will be unduly urged forward by the load.

The force required to move a ton load on the roadway surfaces stated, the corresponding limiting angles of resistances and maximum inclination are embodied in the accompanying table,† the values of the resistances used in its calculation being those given by Sir John Macneil.

---

* The use of "drags" is very *detrimental* to the surface of the road.
† Law and Clark's "Roads and Streets."

Description of the Road.	Force in lb. required to move one ton.	Limiting angle of Resistance.	Maximum Inclination which should be given to the Road.
Well-laid pavement...	31·4	0° 48'	1 in 71
Broken stone surface on a bottom of rough pavement or concrete	44	1° 7½'	1 in 51
Broken stone surface laid on an old flint road	62	1° 35'	1 in 36
Gravel road	140	3° 35'	1 in 16

To avoid considerable expense in maintenance a gradient of 1 in 30 is the steepest that should be adopted; and on inclines exceeding 1 in 20 vehicular traffic ascends only with great difficulty, as shown by the horses oftentimes voluntarily taking up a zigzag course in search of an easier gradient.

Inclines of 1 in 10 to 1 in 8, as on mountain passes or paths, are fit for beasts of burden only, but even steeper gradients than these are to be found in some towns in England, as, for example, at Blackburn,* where the maximum gradient of macadamised roads is 1 in 6·03 and of paved roads 1 in 6·05. In Merthyr Tydfil a 15-ton steam roller was used in making a road with a gradient of 1 in 9½, and, it is stated, did the work well.†

The maximum gradients adopted for road in Central India‡ are:—

For first-class roads ... ... 1 in 25
For second and third class roads ... ... 1 in 20
For fourth-class roads ... ... 1 in 18

ROAD MATERIALS AND CONSTRUCTION.

Materials.—The essential qualities of a good stone or "*road*

---

* "Proceedings of the Association of Municipal and County Engineers," vol. xii., p. 19.
† "Proceedings of the Association of Municipal and County Engineers," vol. iv., p. 97.
‡ "Roorkee Treatise."

metal" are hardness, toughness and power to resist the action of the weather. It should also possess strength to resist compression, and should when laid well unite with its own angles to form a compact surface not easily torn up by traffic, without the aid of much "*binding material*." These qualities are not always to be found together in the same class of stone, but some of the best descriptions for the purpose are found among the igneous rocks, such as granite and trap-rock, or whinstone.

Whilst there are certain general tests which are useful in selecting stones for use as a road metal, yet the only satisfactory means of judging as to its suitability for the purpose is by making an experimental trial for a sufficient length of time upon a roadway with a known amount of traffic. The following preliminary tests are suggested by Mr. Boulnois[*]:—

1. Ascertain from local persons, such as masons, quarrymen and others, their opinion of the qualities of the stones in the neighbourhood.

2. Make a trial of the stone for *toughness* by setting a good stone-breaker to work upon a heap of the stone as quarried and carefully watching how much he can break in an hour.

3. Ascertain what power the stone has to resist abrasion, as by putting the broken metal into a revolving cylinder and then carefully noting by weight what the cubes lose by contact with each other, or press the stone against a grindstone with a uniform pressure and note the loss by such contact.

4. The power to resist compression may be easily ascertained by placing small cubes in a hydraulic press and noting under what pressures each cube will crush.

5. The effect of weather is not easily ascertained artificially, although it is suggested that a good test may be made by soaking the stones in a saturated solution of sulphate of soda, and then on exposure to the air, if soft, it is said the stone will disintegrate as if under the action of thaw succeeding frost.

The specific gravity of a stone is no guide as to its fitness for a road metal.

---

[*] "Municipal and Sanitary Engineers' Hand-Book."

Also it is important to bear in mind that a stone should not be selected merely for its *high resistance to crushing stress*, but that *toughness* is an equally essential property for a good and durable material. This is illustrated by the following results of experiments made by Messrs. D. Kirkaldy & Son*:—

	Crushed, steelyard dropped, lb. per square inch.
Common black *flint*, from chalk-beds near Grays, Essex	32,350
Cherbourg stone (quartzite) containing about 93 per cent. of silica...	31,719
Guernsey granite (good average specimen)	28,525

From the above it will be seen that the universally-admitted best material (Guernsey granite) comes out lowest, and the worst material (flints) gives the highest result—indicating clearly that a high-crushing stress alone is no evidence of the durability of a stone for macadamising purposes.

Local stones not suitable for road making are frequently made use of owing to their cheapness in *first cost*, but, generally speaking, it will be better policy to obtain a superior material from a distance if there is much heavy traffic. Soft material should not be mixed with hard, either for the construction or maintenance of a road, as one will obviously wear faster than the other and thus make the surface very uneven, unpleasant to ride over, and full of hollows which retain the wet and damage the road. The hard metal should be reserved for the surface coating.

A few of the most important materials will now be briefly described.

GRANITE† is a stone of a crystalline granular structure, occurring in large quantities in the older geological formations. It is an igneous rock, and probably is the product of the metamorphism of older sedimentary strata. It is found below the "primary" stratified rock, but occurs in beds, veins and dykes, oftentimes injected from below through and

---
* Vide *The Contract Journal*, November 25, 1891.
† From Latin, *granum* = grain.

over the adjacent strata; also in large masses gradually changing in character and passing into the surrounding rocks.

True granite consists of crystals of quartz and felspar,* mixed with thin plates or scales of mica.

The *quartz* or silica occurs in hard glassy grey or colourless amorphous lumps.

The *felspar* is in the form of opaque crystals, of very irregular size, and of a white, grey, yellowish pink, red or reddish brown colour, which gives the tone to the mass.

The *mica* consists of small glistening scales, capable of being flaked off with the knife, and having a yellow or brownish-yellow, and sometimes dark grey or black, colour.

The durability of the granite depends principally upon the quantity of quartz it contains, upon the nature and regular distribution of the felspar, upon the smallness of the quantity of mica, and upon the absence of iron in any form.

Potash felspar and lime felspar are those kinds occurring most frequently in granite—both kinds being sometimes found in the same stone. The potash felspar is the more liable to decay. "All granites are not suitable for road making. When a granite becomes weathered the felspar may decompose into kaolin or china (porcelain) clay.† The commencement of this alteration is indicated under the microscope by the turbidity of the felspar. At the quarries it is often necessary to reject large quantities of stone for road purposes because of this change. All the toughness is gone out of it, and the quarrymen speak of it as 'dead.'"‡

Mica readily decomposes, and its presence is therefore a source of weakness.

The granite should be close-grained. Large, dull crystals of felspar indicate weakness.

The following list of granite quarries§ indicate the districts in Great Britain and Ireland from which most of the road-making material is obtained.

---

* Also spelt *feldspar*.
† This is the case with the potash felspar, contained in large proportions in the Cornish and Devonshire granites.
‡ "Municipal and Sanitary Engineers' Hand-Book," by Mr. H. Percy Boulnois.
§ Abstracted from "Notes on Building Construction" (Rivington) vol. iii.

Name of Quarry.	Colour of Stone.	Remarks.
ENGLISH.		
Bardon Hill (Leicestershire)	Greenish	Road metal much used in Midlands.
Clee Hill (Shropshire)	—	Used for road metal.
Grooby (Leicestershire)	Pink and green	Syenite. For paving setts and road metal.
Herm (near Guernsey)	Grey	Syenite. Fine-grained, hard and durable; used chiefly for paving; apt to become slippery.
La Moye (Jersey)	Pink and grey	Syenite. For paving setts and road metal.
Markfield (Leicestershire)	Dark green	Do. Do.
Mountsorrel (Leicestershire)	Pinkish brown	Syenite. Hard and durable; used chiefly for paving and road metal.
Portmadoc (Merionethshire)	—	Syenite. For road metal.
SCOTCH.		
Dancing Cairn (Aberdeenshire)	Usually grey; sometimes red	Buildings in Aberdeen. Used in London for curbs, paving, &c.
Tillyfourie (Aberdeenshire)	Bluish grey	Much used in London for paving and setts.
Tyrebagger (Aberdeenshire)	Grey	Do. Do.
IRISH.		
Dalkey (Dublin)	Grey	A good stone for building, road metal and paving setts. Hard to work.
Newry (Down)	Grey	A hard good stone for general building purposes, very durable; one of the best granite in Ireland.

Granite is quarried by wedging for large blocks and by blasting for smaller pieces and road metal.

The *Scotch granites*, especially those from Aberdeen (grey) and Peterhead (red), are noted for their durability and beauty.

Cornish and Devonshire granites are not so good.

Leicestershire granites, generally syenites, are tough, hard and well suited for paving setts and road metal.

Jersey and Guernsey granite (syenitic) is hard, durable, used for paving, but apt to wear slippery.

The hardest and best stones for macadam are Guernsey granite and Penmaenmawr graywacké.* The crushing resistance of the latter is greatly in excess of granite, being equal to about 7¼† tons per square inch. The maximum crushing resistances of inch cubes of various granites vary from 1 to 6 tons.‡

The *crushing resistances* of various granites are shown in the following table§ :—

Locality of Stone.	Pressure per sq. inch.	
	To fracture.	To crush.
Herm ‖	4·77	6·64
Aberdeen (blue) ‖	4·13	4·64
Heytor ‖	3·94	6·19
Dartmoor ‖	3·52	5·48
Peterhead (red) ‖	2·88	4·88
Peterhead (bluish grey) ‖	2·86	4·36
Penrhyn ‖	2·58	3·45
Killiney (grey felspathic) ¶	—	4·81
Ballyknocken (coarse grey) ¶	—	1·43
Ballybeg, Carlow (grey felspathic) ¶	—	3·17
Aberdeen[1]	—	5·16
Mountsorrel[1]	—	5·74
Bonaw, Inverary[1]	—	4·87

* *Graywacké* = a conglomerate of grit rock, consisting of rounded pebbles and sand firmly united together. The term, derived from the *grauwacke* of German miners, was formerly applied in geology to different grits and slates of the Silurian series, but it is now seldom used.— Webster's "International Dictionary."
† Sir W. Fairbairn.
‡ Mr. Mallet.
§ Law and Clark on "Roads and Streets."
‖ The table on *crushing resistance* is based upon data given by Sir John Burgoyne; ¶ Prof. Hull; [1] and Sir W. Fairbairn.

The relative durability of granites, from experiments on stones laid down in 1830 at Limehouse, so as to be exposed to the heavy traffic from the East and West India Docks, are given by Mr. Walker as follows* :—

Name of Stone.	Super. area.	Loss of weight per sq. foot.	Vertical wear.	
	sq. feet.	lb.	inch.	relative.
Guernsey granite	4·73	·95	·060	1·000
Herm granite	5·25	1·05	·075	1·190
Budle whinstone	6·34	1·22	·082	1·316
Peterhead blue granite	3·48	1·80	·131	2·080
Heytor granite	4·31	1·92	·141	2·238
Aberdeen red granite	5·38	2·14	·159	2·524
Dartmoor granite	4·50	2·78	·207	3·285
Aberdeen blue granite	4·82	3·06	·225	3·571

Mr. Walker has also proved that the relative wear of Guernsey and Aberdeen granite is as 1 to 6.

Elvan is a term used in Cornwall and Devon to denote certains veins of felspathic or porphyritic rock, usually of a whitish-brown colour, consisting of quartz and orthoclase (potash felspar). It occurs in dykes and veins proceeding from the granite, but differs from *true* granite in that it has no mica. It varies in texture, and is sometimes of a laminated nature; it is durable and suited for road metal and building purposes.

Gneiss is "a crystalline rock, consisting, like granite, of quartz, felspar and mica, but having these materials, especially the mica, arranged in planes, so that it breaks rather easily into coarse slabs or flags. Hornblende sometimes takes the place of the mica, and it is then called *hornblendic* or syenitic gneiss. Similiar varieties of related rocks are also called gneiss."† In appearance and properties gneiss resembles granite, but is less strong and durable; it is used for ordinary masonry in the neighbourhood where found, and, from its stratified nature, makes a good material for flagstones but not suitable for road metal.

---

* Law and Clark on "Roads and Streets."
† Webster's "International Dictionary."

Syenite and Syenitic Granite.—*Syenite* is named from Syene, in Upper Egypt, where the rock is found.

True syenite consists of quartz, felspar and hornblende.

Syenitic granite consists of quartz, felspar, mica and hornblende.

Hornblende is an important mineral, occurring in great variety in composition and appearance, and is a silicate of magnesia, iron, lime and alumina, and often containing other substances. In *syenite* it takes the place of the mica of ordinary granite, whilst in *syenitic granite* it forms the fourth constituent, added to those of true granite. The commoner varieties of hornblende are dark green and black—these forming a large portion of the mass of greenstone (also known as *trap* or *whinstone*.) The mineral is hard and tough, and crystallises in prisms.

"The syenitic granites are on the whole tougher and more compact than the ordinary granites, take a fine polish and are exceedingly durable."* The darker colours are found to be the most durable.†

"Syenite occurs at Malvern and Charnwood Forest. It appears that both granite and syenite occur at Mountsorrel. The rock is of a pink colour, and has been worked also at Grooby (Leicestershire.) Bardon Hill is composed of it."‡

Trap Rocks.—" Trap "§ is an old term indefinitely applied to any eruptive rock, and is rather loosely used to designate various dark-coloured, heavy *igneous* rocks, including especially the felspathic-augitic rocks, basalt, dolerite, amygdaloid &c., but including also some kinds of diorite.||

Greenstone, also known as *trap* or *whinstone*, consists of felspar and hornblende; it sometimes has a granular crystalline structure, the grains being much finer than in granite, and at times so compact as to be without apparent grains. The

---

* Page's "Practical Geology."

† "Municipal Engineers' Hand-Book," by H. Percy Boulnois.

‡ Woodward's "Geology of England and Wales."

§ *Trap*, Sw. *trapp*; akin to *trappa* = stairs; so called because the rocks of this class often occur in large, tabular masses, rising above one another like steps.

|| Webster's "International Dictionary of the English Language."

colour is usually dark green, but varies. Greenstone is compact, hard, tough and durable, and splits up into small blocks, so that it is well suited for paving setts and road metal, but not for use in large masonry works.

Three varieties in common use may be mentioned—Penmaenmawr, Bardon Hill and Whinstone. *Penmaenmawr* stone, near Caernarvon, is a felstone occurring as an intrusive mass in the Cambrian (Lower Silurian) rocks near Conway.[*] It is easily split by cutting a fine line with an axe in the direction required, and then giving the stone a few smart taps with a hammer.[†] *Bardon Hill stone* (Leicestershire) is largely used in the Midlands as a road metal. *Whinstone* is found in Wigtownshire, near Selkirk, in Kincardineshire, near Haddington, near Edinburgh, at Falkirk, in Perthshire, Fifeshire, Inverness, Ross and other places in Scotland.[‡]

BASALT[3] is "a rock of igneous origin, consisting of augite and triclinic felspar, with grains of magnetic or titanic iron, and also bottle-green particles of olivine frequently disseminated. It is usually of a greenish-black colour, or of some dull-brown shade or black. It constitutes immense beds in some regions, and also occurs in veins or dykes cutting through other rocks. It has often a prismatic structure, as at the Giants' Causeway in Ireland, where the columns are as regular as if the work of art. It is a very tough and heavy rock, and is one of the best materials for macadamising roads"[§]; also it affords great resistance to crushing and is well adapted for crushing, &c.

Rowley Rag, a basalt found at Rowley Regis (Staffordshire), is columnar, and was considered by Jukes to be a lava flow during the cone period. It is used as a road metal, for paving setts, and for the manufacture of artificial stone. The material is also found in Armagh, Antrim and Londonderry.

---

[*] Woodward's "Geology of England and Wales."
[†] Seddon.
[‡] "Notes on Building Construction," vol. iii. (Rivington).
[§] Latin, *basaltes* (an African word), a dark and hard species of marble found in Ethiopia.
[||] Webster's "International Dictionary."

*Great Ayton** stone is largely used in the north of England for repairing macadam roads. It is a basaltic rock obtained from Great Ayton, Cookfield and other places along the north of England dykes. The stone has a specific gravity of about 2·7, and is composed† chiefly of silica, alumina, ferric and ferrous oxide, lime and magnesia.

CLAY SLATE.—Clay slate is a primary stratified rock of considerable hardness and density, but is no use for road making, as it crumbles upon exposure and make a good deal of mud.

LIMESTONE.—The name *limestone* is applied to rocks consisting chiefly of calcium carbonate or carbonate of lime. There is, however, considerable difference both as regards chemical composition and physical characteristics even among stones of the same class. Limestone is a stratified rock, of sedimentary origin, and occurs in geological formations of all ages, the beds oftentimes being of enormous thickness.

Limestones are classified by the engineer according to their physical characteristics, as follows: Marble, compact limestone, granular limestone, shelly limestone and magnesian limestone.

Marble is the name given to any limestone which is sufficiently hard and compact to take a fine polish. It generally consists of pure calcium carbonate. "In the absence of better material marble may be used for road metal and paving setts, but it is brittle and not adapted to withstand a heavy traffic. Roads made with it are greasy in wet weather and dusty when dry."‡

Compact Limestone.—This consists of calcium carbonate, either pure or mixed with sand and clay. It has a dull earthy appearance, and is grayish blue, black or mottled in colour. Some of the carboniferous limestones, the Lias limestone and Kentish rag (cretaceous system) are of the com-

---

* "Proceedings of the Association of Municipal and County Engineers," vol. xiii., p. 98.
† *Quarterly Journal of the Geological Society*, May, 1884.
‡ "Notes on Building Construction," vol. iii.

H

pact class. *Kentish rag* occurs in the Greensand formation, and is found in the central part of Kent, in the neighbourhood of the towns of Sevenoaks, Maidstone, Lenham and others. Ragstone is suitable for paving setts and curbs, and is used for road metal, but makes a dusty road in dry weather. In respect to this stone Mr. Francis J. C. May, M.I.C.E., observes* that "Kentish Rag is one of the oldest and one of the very best of building stones. There are several large quarries in the neighbourhood of Maidstone, which form one of its staple industries. It is a very useful stone for road metal, and is largely used on roads of moderate traffic and where the subsoil is dry. Its use is also valuable on hills in conjunction with flints. Indeed, a road made of rag and flints, in the proportion of two of rag to one of flints, will last longer and wear better than a road made of either material separately."

Compact limestone in Torquay has been found to be "well fitted for forming good smooth roads when not required to carry excessive traffic."†

Granular Limestone.—This "consists of carbonate of lime in grains, which are in general shells or fragments of shells cemented together by some compound of lime, silica and alumina, and often mixed with a greater or less quantity of sand. . . . In many cases it is so soft when first quarried that it can be cut with a knife, and hardens by exposure to the air."‡ This stone affords some of the principal building stones of this country—as the *Oolites*, including Portland, Bath and Caen stone—but is of no use as a road metal.

Quenast is a limestone from Belgium. At first sight it may be mistaken for granite, but it is softer than granite and does not wear so well. In colour it is a brownish grey.

Shelly Limestone consists (a) of small shells cemented together and showing no crystals on fracture, as—*e.g.*, Pur-

---

* "Proceedings of the Association of Municipal and County Engineers," vol. xiv., p. 120.
† "Proceedings of the Association of Municipal and County Engineers," vol. xi., p. 105.
‡ Rankine's "Civil Engineering."

beck stone—(b) of shells breaking with a highly crystalline fracture, as Hopton Wood stone. *Purbeck stone* (Oolitic) is of a brownish-grey colour, is much used for paving, and i durable, but wears slippery. *Hopton Wood* stone (Derbyshire) is a carboniferous limestone, of a grey colour, is fine-grained, compact, weathers well, and is used for paving, steps, &c. It has been used for paving part of Abingdon-street, London.

Magnesian Limestone is composed of carbonates of lime and magnesia, with a small quantity of silica, iron and alumina. "When the magnesia is present in the proportion of one molecule of carbonate of magnesia to one molecule of carbonate of lime the stone is called a *Dolomite*."* Bolsover Moor and Mansfield stone are magnesian limestones. The stone has been very largely used for building purposes, including the Houses of Parliament. "In Britain it is found in the New Red Sandstone formation, immediately above the coal."†

Limestone, as largely obtained from Skipton and Clithero, is used in many Yorkshire and Lancashire towns, but is not in any way equal to granite or basalt for road construction. Its power of resistance to crushing force is much less, and it is very sensitive to atmospheric changes. Its general use is probably accounted for by its cheapness of first cost.‡ Mr. James Hall, borough surveyor of Stockton (1881), states§ that "limestone alone is a very unsuitable material, from its great affinity for water, which causes it in dry weather to crumble to dust. When mixed with whinstone, flints or other compact material it is useful in causing the whole to bind quickly, and is therefore often used in roads of steep gradients, where it is difficult to get the stones bedded." More than one-half the roads in Ireland are repaired with limestone or limestone gravel, and for *light traffic* the blue limestone or

---

* "Notes on Building Construction," vol. iii.
† Rankine's "Civil Engineering."
‡ "Proceedings of the Association of Municipal and County Engineers," vol. xiii.
§ *Ibid*, vol. vii.

limestone gravel, according to Mr. Dorman (county surveyor of Armagh), makes an excellent road. The white or grey limestone, he observes, is very inferior and wears rapidly.* Mr. H. P. Boulnois, in his valuable "Municipal Engineers' Hand-Book," writes on the use of limestone as a road metal as follows: "Many hundreds of miles of roadways in this country are made with limestone; they often make an excellent surface, as they possess a considerable power of binding together, but weather and very heavy traffic affect them considerably, as they all have a strong affinity for water; their very power of thus cementing themselves together causes a quantity of dust in dry and mud in wet weather."

Mr. Molesworth gives the crushing strain per square inch of "compact limestone" as 7,700 lb.; whilst the average of six tests made by D. Kirkaldy & Sons on the "dark limestone," largely sold as road material by the Buxton Lime Firms Company, gives a crushing strain of 19,349 lb. per square inch.

SANDSTONE is a stratified rock, found in every geological formation above the primary rocks, but the best kinds on the whole are those which belong to the coal formation.† In composition it usually consists of grains of quartz, cemented together by silica, carbonates of lime and magnesia, alumina and oxide of iron. The stone in which the cementing material is nearly pure silica is the most durable, and that containing much alumina is the weakest. As a rule the weathering qualities of the stone are entirely dependent upon the cementing material, but where the grains are of carbonate of lime instead of quartz, and the cementing matter is silica, the grains are the first to fail. Sandstone is found of various colours, as white, yellow, brown, red and blue of many shades —usually depending upon the presence of iron. The stone is very largely used for building, flagging, &c. Prof. Rankine observes that it "is in general porous and capable of absorb-

---
* *Ibid*, vol. xviii.
† Prof. Rankine's "Civil Engineering

ing much water; but it is comparatively little injured by moisture, unless when built with its layers set on edge, in which case the expansion of water in freezing between the layers makes them split or 'scale' off from the surface of the stone. When it is built 'on its natural bed' any water which may penetrate between the edges of the layers has room readily to expand or escape."

"Some of the harder sandstones are used for setts and also for road metal, but they are inferior to the tougher materials, and roads metalled with them are muddy in wet and very dusty in dry weather."*

Prof. Anstead, in writing on roads, considered sandstone to be better than limestone, and hard limestone better than slate, while basalts and granites, he further stated, are exceedingly good or exceedingly bad, according to the proportion of alkaline earths (especially soda) which they contain.

Some of the principal varieties of sandstone in general use are:—

Bramley Fall (Leeds), originally a fairly coarse-grained sandstone of the Millstone Grit formation, of considerable strength and durability, and very extensively used for heavy engineering work. The old Bramley Fall quarries have now almost ceased to be worked, but a good deal of similar stone is now sold under the same name although quarried elsewhere.

Yorkshire stone comes from the coal measures and Millstone Grit series and the New Red Sandstone formation. It is used for heavy engineering work, and for flagging and landings; the best stone for the latter use comes from near Bradford and Halifax.†

Scotgate Ash.—Used for landings, steps, setts, paving and building.

Mansfield stone is a siliceous dolomite occurring in the Permian system, between the New Red Sandstone and the carboniferous series. It is an important building stone.

---

* Wray.
† "Notes on Building Construction," vol. iii.

Craigleith stone is a very durable sandstone, consisting of quartz grains, siliceous cementing material and mica.

Brandon Hill stone (Gloucestershire) is largely used for paving the streets of Bristol.

Pennant stone (Fish Ponds, Bristol) is a good and durable sandstone, largely used for paving, &c.

FLINTS.—These are "found in nodules or as pebbles scattered through the chalk strata and in beds of gravel, apparently left after the washing away of the chalk."* The surface-picked flints are superior to quarry flints, and if tough make good roads, but they are too brittle for very heavy traffic. Mr. Ellice Clark, speaking at a meeting of municipal engineers held at Bristol, gave it as his experience that roads repaired with *flints* required no binding material.

GRAVEL.—Gravel consists of small loose stones which have become rounded and worn by the action of water. It is found in alluvial deposits, drift, sea beaches and river beds, both recent and ancient. If free from large quantities of earthy matter and of a flinty nature it may be used for roads of very light traffic, but will require constant attention to maintain a good surface. Owing to the smooth character of the stones it is difficult to roll in, and often may be seen to move forward in a wavelike manner in front of a steam roller, even when considerable quantities of binding material are used. There is no advantage, therefore, in using "double-screened" gravel, as the fine stuff or "*hoggin*" screened out must invariably be again added in the form of a "*binding material*" before the surface can be consolidated. Gravel surfaces, especially if subjected to much traffic, are seriously affected by wet weather and frost; also during the processes of scraping and scavenging large quantities of road material are unavoidably carted away.

The general remarks of Sir H. Parnell in his valuable "Treatise on Roads" will be of interest. He says: "With respect to the subject generally of road materials, it may be observed that the best descriptions consist of basalt, granite,

---

* Prof. Rankine.

quartz, syenite and porphyry rocks.* The *whinstones* found in different parts of the United Kingdom, Guernsey *granite*, Mountsorrel and Hartshill stone of Leicestershire, and the *pebbles* of Shropshire, Staffordshire and Warwickshire, are among the best of the stones now commonly in use. The *schistus stones* will make smooth roads, being of a slatey and argillaceous structure, but are rapidly destroyed when wet by the pressure of wheels, and occasion great expense in scraping and constantly laying on new coatings. *Limestone* is defective in the same respect. It wears rapidly away when wet, and therefore when the traffic is very great it is an expensive material. *Sandstone* is much too weak for the surface of a road; it will never make a hard one, but it is very well adapted to the purpose of a foundation pavement. *Flints* vary very much in quality as a road material. The hardest of them are nearly as good as the best limestone, but the softer kinds are quickly crushed by the wheels of carriages and make heavy and dirty roads. *Gravel*, when it consists of pebbles of the hard sorts of stones, is a good material, particularly when the pebbles are so large as to admit of their being broken; but when it consists of limestone, sandstone or flint it is a very bad one, for it wears so rapidly that the crust of a road made with it always consists of a large portion of the earthy matter to which it is reduced. This prevents the gravel from becoming consolidated, and renders a road made with it extremely defective with respect to that perfect hardness which it ought to have."

"*Coefficients of quality* for various road materials have been obtained by the engineers of the French Administration des Ponts et Chaussées. The quality was assumed to be in inverse proportion to the quantity consumed on a length of road with the same traffic, and measurements of traffic and wear were systematically made to arrive at correct results. These processes requiring great care and considerable time, direct

---

* *Porphyry* is a term used somewhat loosely to designate a rock consisting of a fine-grained base (usually feldspathic) through which crystals, as of feldspar or quartz, are disseminated. There are red, purple and green varieties which are highly esteemed as marbles.—Webster's "International Dictionary."

experiments on resistance to crushing and to rubbing and collision have also been made on 673 samples of road materials of all kinds. The coefficients obtained by these experiments, which were found to agree fairly well with those arrived at by actual wear in the roads, are summarised in the following table. The coefficient 20 is equivalent to 'excellent,' 10 to 'sufficiently good,' and 5 to 'bad.'"*

Material.	Coefficient of Wear.	Coefficient of Crushing.
Basalt	12·5 to 24·2	12·1 to 16 00
Porphyry	14·1 to 22·9	8·3 to 16·3
Gneiss	10·3 to 19	13·4 to 14·8
Granite	7·3 to 18	7·7 to 15·8
Syenite	11·6 to 12·7	12·4 to 13·00
Slag	14·5 to 15·3	7·2 to 11·1
Quartzite	13·8 to 30	12·3 to 21·6
Quartzose sandstone	14·3 to 26·2	9·9 to 16·6
Quartz	12·9 to 17·8	12·3 to 13·2
Silex	9·8 to 21·3	14·2 to 17·6
Chalk flints	3·5 to 16·8	17·8 to 25·5
Limestone	6·6 to 15·7	6·5 to 13·5

The breaking of stone for road metal is either effected by *hand* or by *machine.* Hand-broken stone makes the best roadways, as machines crush the material and do not turn it out in fairly uniform and cubical pieces as required for road coatings. There are also large quantities of small stuff, that separated by the revolving screen being, however, frequently used as "binding." The tendency of stone-breaking machines is to deliver large quantities of material of a very irregular pyramidal form, a difficulty which has not yet been overcome. Long, thin or flaky pieces of stone may pass several times through a machine before being broken fit for road material. The cost of breaking a hard tough stone by machine is about 1s. per ton, whilst the same work performed by manual labour costs from 2s. to 2s. 6d. per ton.† Prof-

---
* "Encyclopædia Britannica," art. "Roads and Streets."
† The actual prices must, of course, vary with the nature of the stone and the value of the labour.

Rankine states that "the stone-breaking machine of Messrs. Blake breaks stone into cubes of about 1¼ in. in the side, with an expenditure of power at the rate of from 1 horse-power to 1½ horse-power for each cubic yard broken per hour." Mr. A. M. Fowler speaks of having found great difficulty in making good roads with a steam roller and stones broken by Blake's machine, as they were not sufficiently cubical, although they did very well for the bottom.*

The wear and tear of stone-breaking machines is very considerable, and, in a case described† by Mr. Arthur Jacob, B.A., M.I.C.E., the actual repairs cost £124 in twelve months, representing 62½ per cent. of the original price of the machine.

"A good stone-breaker will break 2 cubic yards of hard limestone to the ordinary gauge in a day, and some men will break more. Hard siliceous stones and igneous rocks can only be broken at a rate of 1½ or of 1 cubic yard per day. Of some of the toughest, such as Guernsey granite, a man can only break on an average half a cubic yard per day. River gravel, field stones or flints, which are already of a small size, can be broken at the rate of 3 or 4 cubic yards per day."‡

As 55 per cent. of broken road metal is solid, the weight of a cubic yard can be calculated as follows:—

$W \times 27 \times \cdot 55 =$ weight of a cubic yard broken for road metal, W being the weight of a cubic foot of the stone.

Mr. D. Kinnear Clark, M.I.C.E., gives the following rules§:

(a) *To find the area of surface that can be covered by 1 ton of broken granite, when the thickness of the layer is given:* Divide 32 by the thickness of the layer, in inches, *unrolled*; or, divide 24 by the thickness of the layer, in inches, when *rolled*. The quotient is the area in square yards.

(b) *To find the area of surface that can be covered by 1 cubic yard of broken granite, when the thickness of the layer is given:*

---

* "Proceedings of the Association of Municipal and County Engineers," vol. i., p. 167.

† "Proceedings of the Association of Municipal and County Engineers," vol. ii., p. 82.

‡ "The Maintenance of Macadamised Roads," by T. Codrington.

§ *Vide* "Roads and Streets," by Law and Clark.

When the metal is *not rolled*, divide 36 by the thickness in inches; the quotient is the number of square yards that can be covered. When the metal *is rolled*, divide 27 by the final thickness, in inches, to give the required quotient.

The specific gravity of granites vary from 2·60 to 3·00, the volume of *a ton* equals from 12 to 14 cubic feet, and the weight of a cubic yard of solid granite is from 1·93 to 2·25 tons, or about 2 tons on the average. A cubic foot equals about 1½ cwt.

Granite absorbs on an average 10 lb. (a gallon) of water per cubic yard, or about $\frac{1}{150}$th of its weight.

Various opinions have been entertained as to the proper *size* to which stones should be broken for road metal. Ordinarily the stone is reduced, by means of a steel-faced hammer, to pieces approximating to a cubicle shape and weighing not more than 6 oz.,* which is the average weight of a cube of stone measuring 1·6 in. on its side.

Mr. Boulnois states† that "an old method of gauging used to be 'such a size as the stone-breaker could put in his mouth,' but this was a *varying gauge* and unsatisfactory to all persons concerned, and 'to pass all ways through a ring of 2¼ in. internal diameter' is now the size very often adopted."

Telford specified road metal to be broken so that the largest piece should pass through a ring 2½ in. diameter, a mode of gauging which is certainly more convenient in practice than that of weighing, and has therefore become general.

In connection with the subject of the size or weight of materials to be used, the observations of Mr. E. B. Ellice-Clark, M.I.C.E., made at a meeting of municipal engineers at Hanley, in 1886, will be of interest‡ :—

"Some difference of opinion exists as to the sizes to which stone should be reduced for metalling a road. There is a prevailing opinion that all stones should be broken to pass a

---

* Mr. Macadam required his road inspectors to carry a small pair of scales to test the weight of stones to be used upon his roads.

† "Municipal Engineers' Hand-Book."

‡ *Vide* "Proceedings of the Association of Municipal and County Engineers," vol. xii.

gauge of 1½ in. The writer ventures to express the opinion that this is an error. All the hardest stone, like granite, trap, rock, basalt, the Devonshire dolerite and similar rocks, should be broken to a smaller gauge than flints and the hardest limestone, which in their turn should be broken smaller than such materials as Kentish rag and stones of a similar character. The method of specifying the dimensions of stone should be abandoned for the weight test. Macadam says: 'Every piece of stone put on to a road which exceeds 1 in. in any of its dimensions is mischievous,' and in most of his specifications he insists on no stone weighing more than 6 oz. Parnell adopts 2½ in.* for the largest dimensions. To within the past few years the latter size was very generally adopted, irrespective of the quality of the material. It has been the fashion now for upwards of half a century, when repairing roads with granite and the harder rocks, to have the stones broken as uniformly as possible. The results of this are that, though the general surface may be in good repair, the road will be full of small rises and depressions, the surfaces of which are also rough, stones rising abruptly above the general surface of the road. It is this which causes granite macadam roads to be so unsuitable for light-springed vehicles, such as cyclists use. The author has recently been led to investigate the cause of complaints arising from cyclists when travelling over what was apparently a well-kept road, and he has come to the conclusion that it is of as much importance to have stones of different sizes as it is to have a maximum size. The proportion of different sizes requires yet to be determined. So far as his investigations have gone, he gives the following as closely approximating upon the proper proportions of sizes :—

	Maximum Weight.	Minimum Weight.
	oz.	oz.
Granite and similar rocks ...	3½	½
Flints and similar stones ...	5	¾
Limestones and similar stones ...	6	1

* "Treatise on Roads."

One-half of the total quantity to be of the maximum weight one-eighth of the minimum weight, the remaining three-eighths to be composed of stones varying between the maximum and minimum. This brings us to the question of *binding materials*. A road formed of different-sized stones will require no binding materials. In a former paper on this subject, published ten years ago, the author stated his conviction that the 'decadence of modern roads commenced with the using of binding material,' the introduction of which was coincident with the use of stones broken to a uniform size. Longer experience has confirmed this, and, though in practice he is compelled to use materials to bind (?) roads, he does so very sparingly, and only because of the inability to obtain materials broken to various sizes in sufficient quantity. If the demand is, however, generally set up for proportions of different-sized stones, the necessary quantities will soon find their way into the market."

Mr. Thomas Codrington writes* : " The stone for a *new road* should pass a 2½-in. ring ; for *repairs* 2¼-in. or 2-in. ring. . . . Broken road material contains 55 per cent. solid stone to 45 per cent. of void space. Specimens of good road surfaces weigh from 93 per cent. to 95 per cent. of the weight of the solid stone of which they are made. In the coating of a well-maintained road the proportion of stones of various sizes varies, but generally from one-third to one-half is found to consist of detritus under ¾-in. in diameter, and there is a very constant proportion of about one-fifth of mud and detritus under $\frac{1}{30}$ in. in diameter. This appears to be the amount necessary to fill the voids between the fragments of stone when compacted together. In an ill-kept road, from which the mud is not removed, the proportion of detritus is much higher, and mud may constitute nearly one-half of the coating. In proportion, as the detritus and mud are kept down to the minimum by constant removal from the surface, so will the road be able to resist the action of wet and frost and the wear of the traffic."

---

* *Vide* " Encyclopædia Britannica," art. " Roads and Streets."

CONSTRUCTION.

Generally speaking, there are three classes of macadamised roads—*viz.*,—

(1) Roads as constructed by Mr. Macadam, consisting simply of a coating of broken stone laid upon the natural ground.

(2) Roads as constructed by Mr. Telford, with the distinguishing feature of a "*pitched foundation*" upon which to lay the "*metalling.*"

(3) Roads having a *concrete foundation*, as used by Sir J. Macneil on the Highgate Archway (London) road.

The class of road, quality and thickness of materials to be adopted in any particular situation, will necessarily depend upon the circumstances of the case, and the amount and nature of the traffic accommodated must also be kept prominently in mind in determining upon the details of its construction. There can be no model type of road which can be fitly adopted universally. For a park or pleasure garden it will generally be sufficient to form the roads or drives by simply shaping the earth to a curved contour and covering it with 3 in. or 4 in. of fine gravel, hand rolled; for an ordinary country road, with a comparatively small amount of traffic, a coating of about 9 in. of broken stone, laid upon the natural ground properly shaped, will be adequate; whilst for main roads between large towns and all roads having a considerable amount of traffic a solid pitched foundation with a good coating of about 9 in. of broken granite will be necessary, in order that the road may be as hard and durable as possible, so that loads may be conveyed over it with a minimum expenditure of traction power, which should be the main object to be aimed at in all road-making operations.

Macadam's Plan of Construction.—Although considerable credit is due to Mr. Macadam for the general improvements he effected in roads under his charge by means of the adoption of the practice of putting "broken stone upon a road which should unite by its own angles so as to form a solid hard surface," yet he appears to have entertained some very erroneous notions in respect to the proper construction of *new roads.* He has stated

"That a foundation of bottoming of large stones is unnecessary and injurious on any kind of subsoil."

"That the maximum strength or depth of metal requisite for any road is only 10 in."

"That the duration only, and not the condition, of a road depends upon the quality and nature of the material used."

"That free stone will make as good a road as any other kind of stone."

"That it is no matter whether the substratum be soft or hard." In fact, that he "should rather prefer a soft one to a hard one," or even a bog, "if it was not such a bog as would not allow a man to walk over it."

These ideas are, of course, entirely at variance with the first principles of science and with universal experience. It was thought that when the road materials rested upon a soft and yielding bed they were less likely to be crushed by the passage of heavy traffic over them than when the foundation was hard and solid.

The *foundation*, however, of a road must be looked upon as its most essential part, and "the outer surface (or broken stone coating) should be regarded merely as a covering to protect the *actual working road beneath*, which should be sufficiently firm and substantial to support the whole of the traffic to be carrried over it."[*] In short, the function of the road metalling is to take the *wear*, and that of the foundation to carry the *weight*, of the traffic.

In forming roads wholly of broken stone the ground must be first prepared by levelling its irregularities and forming the surface to the intended contour of the finished roadway. Should the subsoil be of a soft, wet and retentive nature, it must be efficiently drained by means of cross drains, laid in the manner already described, at frequent intervals across the roadway, and discharging into deep side ditches, to be cut on each side of the road to a sufficient depth below the formation surface. In the case of a very soft or boggy ground, layers of faggots or brushwood are also frequently laid with

---

[*] "The Construction of Roads and Streets," by Law and Clark.

advantage over its surface, and upon these the materials are spread. If the tract over which the road is to be formed consists of newly-made ground, as in the case of embankments, it should first be consolidated by punning or rolling. The ground surface having been properly prepared, the broken stone is spread uniformly over the road, by means of a shovel and rake, in successive layers of about 3 in. in depth, allowing, in the absence of a steam roller, time between the layers for their proper consolidation by the traffic. The total thickness of coating which will be required will necessarily depend upon the nature and extent of the traffic and upon the firmness or otherwise of the bed upon which it is laid. From 6 in. to 12 in. is the usual thickness, varying according to circumstances. Over the stone when laid "*binding material,*" consisting of fine gravel, sand or road scrapings, is frequently spread, with a view of assisting the consolidation of the road metal and of making the traffic easier over the new surface. Its use was strongly objected to by Macdam, and although at the present time it is very generally used by surveyors, it should be dispensed with wherever possible, and only used in the least possible quantities at any time. Telford usually covered a newly-metalled surface with about 1½ in. of gravel as a "binding," but this is now rendered unnecessary, the use of the steam roller for effectually and quickly consolidating fresh-laid road metal having become so general.

In respect to this class of road Sir H. Parnell observes that "experience has fully established their unfitness for roads of great traffic in comparison with those made with a proper foundation. The reason is very obvious, for if a coating of small broken stones be laid on the natural soil the weight of carriage wheels passing over it forces the lower course of the stones into the soil, while the soil is forced up into the interstices between them; the clean body of stones first laid on to make the road is thus converted into a mixed body of stones and earth, and, consequently, the surface of the road cannot but be very imperfect as to hardness. It is necessarily heavy in wet weather, on account of the mud the earth makes on its surface, and, in warm weather, on account of a quantity of dry dirt. A road made on this plan will require, for two or three years after it is said to be finished, the expenditure of large sums in new materials to bring it into anything like even an imperfectly consolidated state; and, after all that can be done, such a road will always run heavy, and break up after several frosts; for, as the natural soil on which such a road is laid is always more or less damp and wet, it will necessarily keep the body of materials of which the road is made damp and wet; in consequence of which the surface of the road will wear down quickly. Hard frosts will penetrate through the materials into the under soil, and when thaws take place break up the whole surface."

Although this form of construction has been very largely adopted throughout the country, it is only suitable for minor roads with a small amount of traffic.

Telford's Plan of Construction.—The most prominent feature of this mode of construction, as has already been said, is the use of a solid "*pitched foundation.*" A specification of Telford's, describing the manner of forming a road of this kind of 30 ft. in width, taken from a contract for making a part of the Holyhead road, has been given at length.

The method of laying down the foundation, also the objects and merits justly claimed for it, are best set forth in Mr. Telford's own words*: "This foundation is a regular close pavement of stones, carefully set by hand, and varying in height from 8 in. to 6 in., to suit the curvature of the road; these stones are all set on edge, but with the flat one lowest, so that each shall rest perfectly firm. The interstices are then pinned with small stones, and care is taken that no stone shall be broader than 4 in. or 5 in., as the upper stratum does not bind upon them so well when they much exceed that breadth. The pavement thus constructed is quite firm and immovable, and forms a complete separation between the top stratum of broken stones and the retentive soil below. Any water which may percolate through the

* First report of Mr. Telford on the Holyhead road, May, 1824; also sixth report, May 23, 1829.

surface is received among the stones of the pavement, and runs from them into the next leading or cross drain, and there escapes."

"The different parts of the Holyhead road, which have been newly made with a strong bottoming of stone pavement, place beyond all question the advantage of this mode of construction; the strength and hardness of the surface admit of carriages being drawn over it with the least possible distress to horses. The surface materials, by being on a dry bed and not mixed with the subsoil, become perfectly fastened together in a solid mass, and receive no other injury by carriages passing over them than the mere perpendicular pressure of the wheels; whereas, when the materials lie on earth, the earth that necessarily mixes with them is affected by wet and frost; the mass is always more or less loose, and the passing of carriages produces motion among all the pieces of stone, which, causing their rubbing together, wears them on all sides, and hence the more rapid decay of them when thus laid on earth than when laid on a bottoming of rough stone pavement. As the materials wear out less rapidly on such a road the expense of keeping it in repair is proportionally reduced. The expense of scraping and removing the drift is not only diminished, but with Hartshill stone, Guernsey granite, or other stone equally hard, is nearly altogether avoided."

In a roadway of 30 ft. in width, constructed after Telford's plan, a convexity or rise of 6 in. was obtained at the crown, the convexity of 4 ft. from the centre being $\frac{1}{2}$ in., at 9 ft., 2 in., and at 15 ft., 6 in., thus giving the form of a flat ellipse.

The stone employed for the foundation may be such as would not be suitable for other purposes, as metalling, building retaining walls, &c., and may therefore frequently be obtained locally for this purpose. Chalk may be used to form the foundation if kept sufficiently deep to be entirely out of the reach of frost. Mr. M'Neill states[*] that "sandstone, limestone or schistus, or such as can be had in the neighbourhood, may be used for the purpose; any stone, almost, will

---

[*] Evidence of Mr. M'Neill before a Select Committee of the House of Commons in May, 1830.

answer that will bear weight and not decompose by the atmosphere."

The " bottoming " being thus properly performed, a coating, of about 6 in. or 9 in. in depth, of some hard and tough road metal, such as Guernsey granite or whinstone, should then be uniformly laid down, and rolled in in layers by the aid of the steam roller or consolidated by the traffic, the surface being well attended to during the process, and all hollows or ruts which may appear being at once filled in with road metal. If the road material is, as it should be, in angular cubical pieces, so that their angles may interlock with each other, no binding material will be required.

In and about the metropolis the *foundation* of a macadamised road is usually formed of "*hard core*," a term which is applied to a heterogeneous mixture consisting of broken stone, brick rubbish, clinker, broken pottery and various other hard materials. In the north of England, and generally in towns situated near blast furnaces, the foundation is formed of *slag*. The thickness of the hard-core foundation will depend upon the nature of the subsoil, upon the amount of traffic, and also upon the nature of the material used, but 12 in. may be regarded as the minimum thickness, and this should be consolidated by rolling to about 9 in., all hollow places being at once filled in and made level.

Upon this a 5-in. layer of Thames ballast or gravel should be uniformly spread and consolidated to about 3 in. in depth. Then, to receive and withstand the wear and tear of the traffic, a 6-in. coating of broken Guernsey granite should be laid down in two layers and well rolled. About ½ in. of sharp sand is frequently scattered over the surface, and the whole consolidated by rolling and watering. In London a road constructed after the above manner would cost about 6s. per square yard.

For ordinary *country roads* the ground should be excavated in the usual manner to an approximate circular segment, and the foundation formed of rough stones, flints, &c., to a thickness of at least 12 in., and covered with a 6-in. coating of broken flints. Binding material is not generally used, and

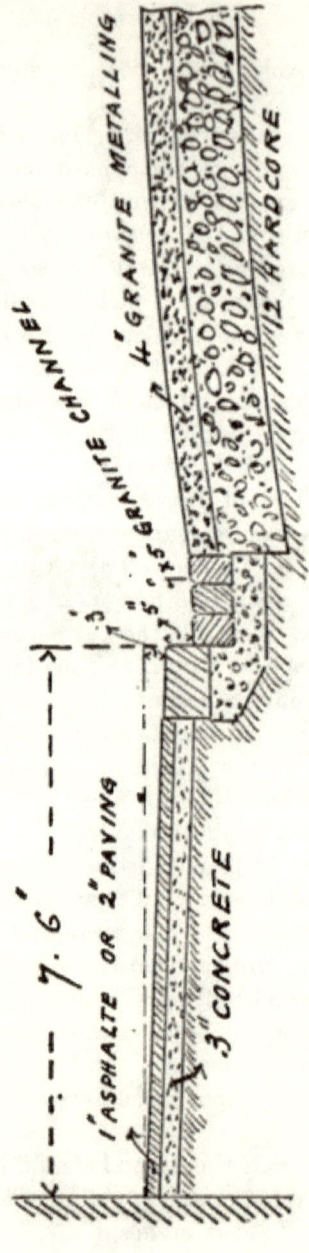

Fig. 26.—Residential Road.

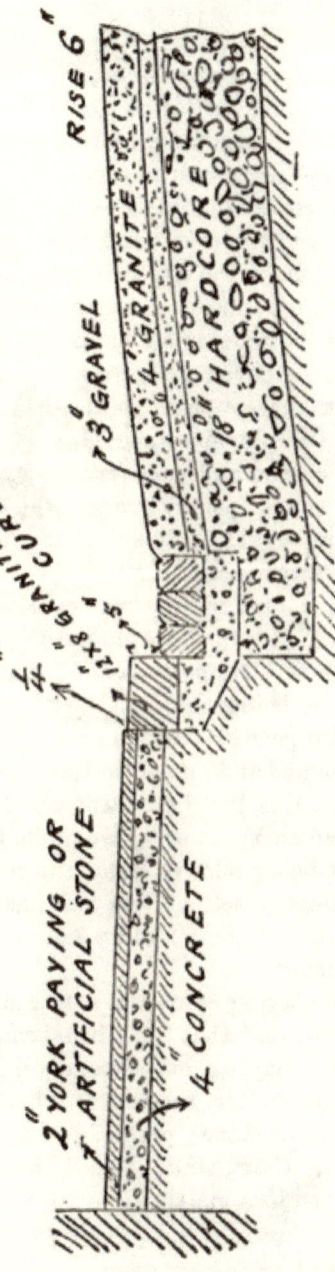

Fig. 27.—Business Road.

steam rolling in many country districts is almost unknown, the new metalling being left to be gradually worked in by the traffic, assisted by occasional attention to the raking in of ruts and hollows appearing on the surface.

Roads with Concrete Foundations.—Roads have been constructed with *concrete foundations* and have proved very successful, but the process is too expensive for general application, and has therefore not been very extensively adopted, except in special situations, upon bad ground, or where the road is intended to be paved with granite, wood or asphalte. Mr. Henry Law, c.e., states* that "one of the principal advantages attending the employment of concrete as a foundation for roads is that a good and solid road may be made with materials such as round pebbly gravel, which, on any other mode of application, would be ill-suited to the purpose, and would form a very imperfect road."

The concrete for the foundation may consist of good clean gravel, containing a little sand, mixed with hydraulic lime in the proportion of four or six parts of the former to one of the latter. The mixture being thoroughly well incorporated, is at once wheeled into position, spread over the roadway to a depth of 6 in., trimmed and spread to the form of the roadway surface. Upon this, just before the concrete has become hard, a layer of broken stone or gravel, 3 in. in thickness, is laid, the traffic not being allowed upon the surface until the foundation is thoroughly set. When this has taken place a top covering of broken stone is laid on 3 in. in depth, forming a solid road throughout.

In respect to the laying on of the first coat of stone, Mr. Thomas Hughes observes† that "the beneficial effect arising from the practice of laying on the gravel exactly at the proper time—*i.e.*, just before the concrete has become quite hard—is that the lower stones, pressed by their own weight and by those above them, sink partially into the concrete, and thus remain fixed in a matrix from which they could not

---

* "Roads and Streets," by Law and Clark.
† *Vide* "The Practice of Making and Repairing Roads."

easily be dislodged. The lower pebbles being thus fixed, and their rolling motion consequently prevented, an immediate tendency to bind is communicated to the rest of the material; a fact which must be evident if we consider that the state called binding, or rather that produced by the *binding*, is nothing more than the solidity arising from the complete fixing and wedging of every part of the covering, so that the pebbles no longer possess the power of moving about and rubbing against each other."

This form of construction was applied by Mr. Charles Penfold to the Brixton and Walworth roads with marked success, also in Southwark-street a 12-in. bed of lias lime concrete was put down, and Mr. M'Neill suggested this plan for use upon the Highgate Archway-road. The circumstances leading to its adoption upon the latter road are interesting as well as instructive, owing to the difficulties to be overcome, and will therefore be briefly described.

The Highgate Archway-road, about 1½ miles in length, was originally made by a private company, at great expense, owing to the nature of the subsoil, which consisted of sand, clay and gravel. An unsuccessful attempt having been made to form a tunnel through the hill, open cutting was resorted to, and the roadway made by laying large quantities of gravel and sand upon the natural soil and thickly coating the same with broken flint and gravel. The result, however, not being satisfactory, the road was discarded by much of the traffic, and the company, being naturally anxious to improve its condition, among other schemes took up the road material and covered the subsoil with pieces of waste tin, upon which were placed gravel, flints and broken stone, but without attaining the desired success.

The road continued in a very bad state for some considerable time, as may be seen from Mr. Telford's annual reports to the Parliamentary Commissioners, until in 1829 an arrangement was made by the commissioners with the Highgate Archway Company for taking the road under their management; whereupon it was re-formed and put into proper repair. "In order to accomplish this," writes Sir Henry

Parnell,* "several experiments were tried, by draining the surface and subsoil, and by laying on a thick coating of broken granite; but from the wet and elastic nature of the subsoil the hardest stones were rapidly worn away by the wheels of carriages, but much more by the friction of the stones themselves against each other; for in a very short time they were found to become as round and as smooth as gravel pebbles, even at the bottom of the whole mass of road materials." It was therefore evident that to form a perfect road, which might be kept in repair at a moderate expense, it was necessary to establish a dry and solid foundation for the surface broken stones; but as no stones could be obtained for making a foundation of pavement but at a very great expense, a composition of Roman cement and gravel was suggested by Mr. M'Neill, and this on a trial was found to answer effectually.

The foundation was formed of concrete, consisting of Roman cement, washed gravel and sand, mixed in the proportion of one cement to eight of gravel. In forming the bed of the road "there were four drains formed longitudinally, and there were secondary drains running from those to the side channel drains, and those again to drains outside the footpaths, covered with brick, and they all communicated with each other, and discharged the water into proper outlets."† The cross drains occurred at intervals of 30 yards, and the intermediate small drains every 10 yards under the cement. Mr. M'Neill also stated that this special drainage was rendered necessary through the nature of the ground— the road being cut through a clay soil, with high banks on each side, so that all the surface water descending from the slopes and Highgate Hill came down and rested in the hollows of the subsoil.

"On the prepared centre of 6 yards in width, after it had been properly levelled, the cement was laid on, mixing it first in a box with water, gravel and sand in certain proportions "†

---

* *Vide* "A Treatise on Roads."

† Evidence of Mr. M'Neill before a Select Committee of the House of Commons, in May, 1830.

as above. The concrete, when in position, was found to have set in a quarter of an hour, and " in about four minutes after being laid a triangular piece of wood, sheeted with iron, was indented into it, so as to leave a track or channel at every 4 in. for the broken stones to lie and fasten in."* " This triangular indent had an inclination of fully 2 in. from the centre to the sides; so that if water came through the broken stones it ran off the cemented mass into the longitudinal drains."

Quite recently some concrete blocks about 12 in. by 6 in. by 4 in. were met with in the course of excavations in this road. They occurred at a depth of about 2 ft. 6 in., and are now to be seen in the museum of the Hornsey District Council—the authority now responsible for the maintenance of that portion of the road north of the "Arch."† These blocks and slabs, and also faggots, it is stated by a very old resident, were used in the foundations upon soft and treacherous ground.

The foundation stone of the "Arch,"‡ was laid on October 31, 1812, but it was not until the early part of 1830 that Mr. McNeill had completed the laying-in of his concrete foundation and other improvements, and the road thus brought into a satisfactory condition.

The road appears to have been liberally encumbered with " toll gates " from its formation to within a comparatively recent date. The old toll gate in the year 1825 was in the line with the then "Archway" tavern, but some years later was shifted northwards to a point between the "Archway" tavern and the Archway, just north of the Whittington almshouses. This gate finally disappeared in the year 1879 or 1880. The amount of the toll levied was 6d. for horses and 1d. for foot passengers. Another gate crossed the Archway-road, near the "Woodman" public-house, just north of Southwood-lane, and was removed some twenty-five years ago.

---

* Evidence of Mr. M'Neill before a Select Committee of the House of Commons, in May, 1830.

† The portion below the Arch is in the Islington district.

‡ The Arch was of itself a substantial-looking stone structure, but disfigured by a number of ugly brick arches, built across it for the purpose of conveying the Hornsey-lane over the Archway-road cutting.

Within the Hornsey district, opposite the Manor farm, a further gate existed, which, however, was removed in 1863.

As regards the *present experience* of the maintenance of this important highway, Mr. E. J. Lovegrove, A.M.I.C.E., engineer and surveyor to the Hornsey District Council, states that "speaking of the road generally, it is a difficult one to keep in repair, having regard to the peculiar class of traffic and also to the fact that there is a considerable amount of movement in the road itself."*

In an excellent paper, read at Blaydon-on-Tyne before the Association of Municipal and County Engineers, Mr. James Hall observes that "the only objection to a road formed with a concrete bottoming is its first cost, which precludes its general use. In towns, however, where it is considered desirable to have a broken-stone road, a better foundation could not be obtained; one of its great disadvantages is that, becoming a compact mass after setting, the weight is evenly distributed over the whole surface of the roadway, thus preventing, to a very large extent, an uneven surface, which is often found in roads where the traffic is heavy. Concrete can be made both with lime, lias lime, and cement; the latter is preferable. When a concrete foundation is to be used, great judgment should be exercised in the proper time the metal should be spread. Some engineers say that the concrete must be allowed to get thoroughly set, then covered with a thin coating of fine gravel (pit gravel preferred), and the required thickness of metal then spread; while others consider it better to spead the metal upon a thin coating of gravel before the concrete is set, and roll the surface until the metal has partly embedded itself in the concrete. I am strongly of opinion that the latter plan is the better. The road, however, must not be open for traffic until the concrete is quite hard, and until after the first coating of metal has been covered with a thin coat of finely-broken material."†

---

* I am also indebted to Mr. Lovegrove for the above particulars in reference to the toll gates and concrete blocks.

† "Proceedings of the Association of Municipal and County Engineers," vol. vii.

The above *thicknesses of foundation and metalling* of broken-stone roads have also been suggested by Mr. Hall.

	Pinned Foundations.		
	Pinning.	Covering.	Metal.
Country roads...	6 in.	3 in.	4 in.
Suburban roads	9 in.	3 in.	5 in.
Town streets ...	9 in.	6 in.	5 in.

	Broken Stones.		Concrete.	
	Under.	Upper.	Concrete.	Metal, &c.
Suburban roads	9 in.	4 in.	4 in.	3 in.
Town streets ...	9 in.	6 in.	6 in.	5 in.
Country roads...	15 in.	6 in.	10 in.	5 in.

Construction of French Roads.—Large boulders, 8 in. or 9 in. deep, are placed at the sides, as in Fig. 28, and a stone bottoming about 6 in. in depth is formed of rubble stone, upon which comes about 3 in. of small stone packing filling the interstices of the pitched foundation, making together a depth of about 7½ in. This is covered with a coating of hard metalling, broken to pass 1¼-in. ring, and well rolled in with sand as a "*binding*," aided by watering.

CONCRETE MACADAM.

This was introduced by Mr. Joseph Mitchell as an improvement upon ordinary macadamised surfaces. He laid down a concrete macadamised road in Edinburgh of the following proportions: Portland cement, one; broken stones, four; sand, one and a quarter.

"The stones were well screened and watered, and the whole turned over, thoroughly mixed, and spread to a thickness of 3 in. to 5 in., and allowed to harden, after which a second layer was spread, and so on, until the required thick-

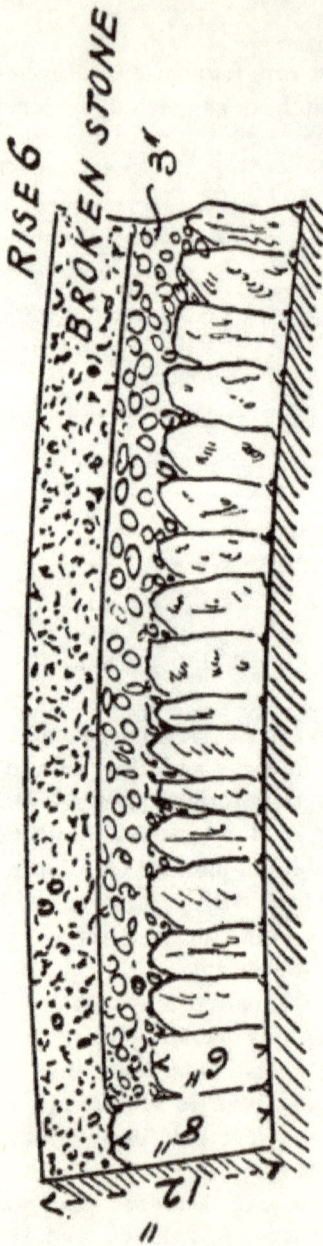

Fig. 28.—Cross-Section of French Road.

ness was obtained. A road thus formed need only be about one-half to two-thirds as thick as ordinary broken-stone roads. The great objection to roads thus formed is, that when the surface becomes worn the cost for repairing them will be considerably more than that of an ordinary road." *
Also, where streets are liable to be frequently opened for sewer, gas, water or other pipe tracks, the trenching will be expensive and repairs difficult.

Mr. H. U. McKie, A.M.I.C.E., as a result of his inspection of the concrete macadam streets of Edinburgh in 1883, stated† that "the streets paved with concrete macadam were in fair order. The pavement is non-absorbent and could be readily washed quite clean, and the scavenging of such streets is brought to the minimum. The cement concrete macadam varied from 9 in. to 12 in. deep. Two modes of forming the streets have been adopted, namely :—

"(a) After the street has been excavated to the proper depth and properly formed, and the ground thoroughly beat down, the first coat is laid down with 2-in. macadam, and blinded and made compact to form required.

" The top coat is laid down with clean-riddled 2-in. macadam, which is grouted with carefully-prepared cement grout, varying in strength from one of cement to three of sharp sand, to one of cement and one of sand, carefully formed to the proper contour of the street.

" (b) The second method is to make a concrete of the broken stone and cement, sand and gravel, and lay it on the road in sections in such lengths that each section can be completed before the cement sets. All the broken stones used are of hard metal, whinstone, granite or porphyry, all hand-broken.

"This class of pavement requires great care in construction to make it a success and to prevent unequal settlement in the street after it is completed. It makes a good street, and the cost varies from 6s. to 9s. per square yard."

---

* "Proceedings of the Association of Municipal and County Engineers," vol. vii. p. 30.
† "Proceedings of the Association of Municipal and County Engineers," vol. x.

Tar Macadam.

Tar macadamised or bituminous concrete roadways are adopted in many towns, and for light traffic are found to be a good substitute for ordinary macadam roads, being impervious to moisture, noiseless,* more sanitary, and less muddy and dusty than macadam. Also, the cost of watering and scavenging is less, as well as the annual cost of maintenance.

This class of road is well adapted for use in the quiet quarters of large towns, in residential areas and the less busy thoroughfares of suburban districts. Mr. Deacon considered tar macadam roadways only suitable for roads having less traffic than 40,000 tons per yard of width per annum and macadam for roads with less than 30,000 tons per yard of width per annum.

The work should always be constructed in the spring or winter months, as if done in the summer the heat of the sun draws the tar out of the pavement. Dry weather is essential, and considerable care and experience will be necessary for successful construction.

The precise methods of laying tar paving for footways and tar macadam for roadways necessarily vary somewhat in different districts, according to local circumstances, but the general method is as follows † :—

The material used (in Leicester) is the blast-furnace cinder and limestone. For *footways* it is sorted by screening and sifting into four different sizes of $1\frac{1}{4}$-in., $\frac{3}{4}$-in., $\frac{1}{2}$-in. and $\frac{1}{4}$-in. gauges, and for roadways to $2\frac{1}{2}$-in., $1\frac{1}{2}$-in. and $\frac{3}{4}$-in. gauges. The cinder and limestone are both heated on an iron floor under which the flues from a fire run. The material is then mixed in its heated state with a sufficient quantity of a mixture (also in a heated state) of pitch, tar and creosote, boiled until they form a tough and thick consistency,‡ when it is, after lying for a few days in heaps, ready for use. The quantity of tar, pitch, &c., depends on the qualities of these

---

* Tar pavement has been called the "silent macadam" in Leicester.
† As adopted by Mr. John Gordon, M.INST.C.E., at Leicester—*vide* "Proceedings of the Association of Municipal and County Engineers," vol. x., from which the following particulars are taken.
‡ The ingredients are boiled together until bubbles rise to the surface and in bursting emit puffs of brown smoke.

articles, and particularly of the tar, which varies very frequently. The quantities are also regulated to the character of the traffic expected on the roadway or footway on which the material is to be used, but the average quantities may be taken to be as follows: 12 gallons of tar, ½ cwt. of pitch, 2 gallons of creosote, 1 ton of stone.

The work is laid down in either two or three layers. For *two-coat footway work* 1-in. material is used for the bottom layer 1 in. thick, and ½-in. material for the top 1 in. thick, making 2½-in. total thickness.

In *three-coat footway work* 1¼-in. material is used, 1½ in. thick for the first or bottom layer, ¾ in. for the second 1 in. thick, and ¼ in. for the top covering ½ in. thick, with the addition of Derbyshire spar sprinkled on the top covering in both cases, to give it a white or variegated appearance, making 3 in. in all. Each layer of material is separately well rolled with hand rollers weighing 13 cwt. each.

The *tar macadam for roadways* is laid down in three layers, the preparation of the material being the same as in the case of footways. The road bed and foundations are prepared similarly as for an ordinary macadamised roadway, and the tarred macadam is laid as follows: The first layer is 3 in. thick and of 2½-in. gauge, the second layer is of 1½-in. gauge and 2 in. thick, and the top coat is 1 in. thick of ¾-in. gauge, and covered with cinder dust, sharp sand or grit, each layer being well rolled with a 15-ton steam roller.

As to the durability of this class of work, Mr. Gordon considered that the life of the roads may be fairly taken at about three years, at the end of which period they will in all probability require *topping*—that is, tracking over and recovering with the material used for the top coat—at a cost of from 9d. to 1s. per square yard.

The *cost of the tar paving*, laid as above described, exclusive of preparing the ground, for three-coat work averaged 1s. 6¼d. per square yard and for two-coat work 11d. per square yard.

The *cost of tar macadam* for roadways was from 2s. 6d. to 3s. per square yard, exclusive of preparation of the ground on special formation.

In old paved streets it only involves the removal of the stones and some few inches of ground, if the foundation be already good, or possibly the carting of a few loads of dry material to make up any apparent soft places in the foundation.

In other cases, where a street is being formed on maiden ground, the formation of a foundation with from 9 in. to 12 in. of what is locally called "rammel"* is necessary, and the covering of the same with furnace ashes, gravel or other suitable material to receive the tar macadam. This, with the necessary excavation, adds from one-third to one-sixth per square yard to the cost, but is requisite even for ordinary macadam roads. The cost of preparing the ground, where no such foundation is required, ranges from 3d. to 6d., the breaking up of the old macadam roads being the most costly; but to set against this there is the value of the old materials.

One important feature, Mr. Gordon further stated, is the *rolling* of the foundations before laying down the first layer of the tar macadam, and he considered a steam roller, of about 10 tons weight, to be necessary for success with such roadways.

Tar macadam is also sometimes laid by merely spreading a coating of broken stone, and, after consolidating it by means of a roller, a mixture of coal tar, pitch and creosote oil is poured over it, and upon this a layer of small stone is put down and well rolled in, the surface being finished as before with stone chippings and rolling.

The system of forming tar macadam roadways in Croydon is described† as follows by Mr. J. Walker, the borough engineer and surveyor :—

"The surface of the road was taken off to allow of an 8-in. coat of tar macadam. The bottom 5 in. consisted of the best of the old road material taken off, after it had been sifted, baked, and, while hot, well tarred. With the gás tar a little

---

* The refuse of the granite quarries.

† *Vide* "Proceedings of the Association of Municipal and County Engineers," vol. x.

pitch was mixed, well boiled and used hot. The stones were turned over until all were well blacked. The material was then taken back to the road, laid on and well rolled. The top 3 in. were Kentish rag, baked and tarred similarly to the bottom coat, and well rolled with a heavy hand-roller; a day or two afterwards it was well rolled with the 10-ton steam roller. A little fine Kentish rag was used to bind it, having been previously baked and tarred as the other. The full cost was about 3s. 6d. per square yard. The old foundation was not disturbed, but if the foundation of a new road and new materials were required the cost would not be under 4s. per yard super."

The stones which have been used for tar macadam are granite, basalt, mountain limestone, Kentish rag and blast furnace slag. Hard limestone has been found to be preferable to the siliceous or igneous rocks, as it wears more evenly.

COST OF CONSTRUCTION OF MACADAMISED ROADWAYS.

The cost of constructing a macadamised roadway varies very widely, according to the locality, facility for obtaining road materials, the price of labour, and many other considerations.

For a first-class macadam roadway the following is a brief specification: Excavate to a depth of 16 in. below finished level of proposed road surface; level and properly consolidate

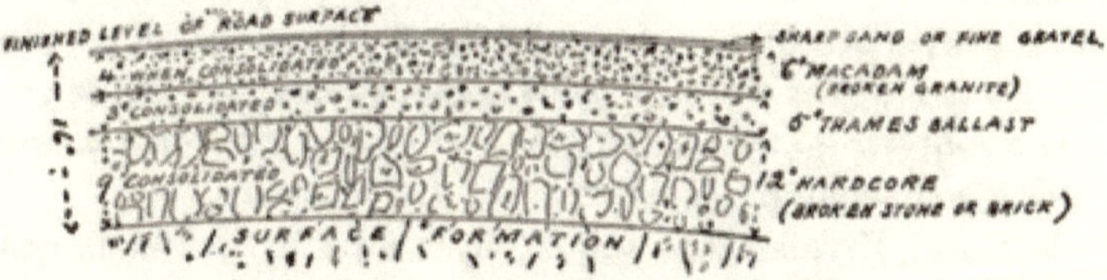

Fig. 29.—Cross-Section of a First-Class Macadam Road.

surface formation to the required contour; lay in a bed of "hard core" of broken stone or brick 12 in. in thickness, and consolidated to 9 in. by rolling with a 10-ton steam roller, and make up all hollow places; spread a layer of Thames ballast or gravel, 5 in. thick, consolidated to 3 in. by rolling; lay down two 3 in. layers of 2-in. blue Guernsey granite, and roll in to a finished thickness of 4 in., sharp sand or fine gravel to be spread during the process and the surface well watered (see Fig. 29).

In London a road constructed after the above manner would cost about 6s. per square yard. A lighter section (Fig. 30) suitable for suburban roads, and consisting of 9 in.

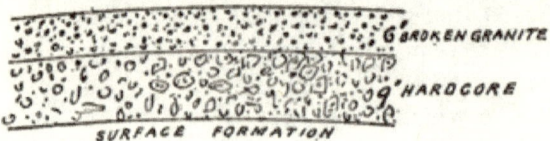

Fig. 30.—Suburban Macadam Road.

of hard core covered with a 6-in. layer of broken granite and well rolled, would cost about 3s. 6d. per square yard.

The following are the approximate prices of road materials in the vicinity of London :—

		Per ton.
Blue Guernsey granite, broken to pass a 2-in. ring,	13s. 10d.
,, ,, ,, ,, ,, 1½-in. ,,	14s. 4d.
Alderney granite ,, ,, 2-in. ,,	14s. 0d.
, ,, ,, ,, 1½-in. ,,	14s. 6d.
Quenast* (limestone) ,, ,, 2-in. ,,	13s. 0d.
,, ,, ,, 1½-in. ,,	13s. 6d.

Flints (best Kentish chalk), from 7s. 6d. to 9s. per cubic yard. Double-screened ballast, from 5s. 6d. to 6s. 6d. per cubic yard. Hard core (broken stone or brick), about 2s. 6d. per cubic yard.

The following are the particulars of the cost of materials used in road-making in Plymouth† (1895) :—

Road metal, Elvan stone, unbroken, 3s. 5d. and 3s. 10d. per ton
,, ,, broken to pass through a 2-in. ring, 5s. and 5s. 5d. per ton.

---

* *Quenast* is a limestone from Belgium, of brownish grey colour. At first sight looks like granite, but is softer, and does not wear so well.

† "Proceedings of the Association of Municipal and County Engineers," vol. xxi.

J

Limestone, for tar paving, 3s. per ton.
First-class granite setts, 3 in. by 6 in. by 6 in. to 8 in. long,
22s. per ton.
„   „   3 in. to 5 in. by 6 in. to 8 in. long,
23s. per ton.
Second-class „   3 in. to 4 in. by 6 in. by 9 in. long,
18s. 9d. per ton.
Granite flagging, 10s. per square yard.
Granite curb, bevelled, 12 in. by 8 in., 1s. 2d. per lineal foot.
„   „   12 in. by 8 in., circular on plan, 1s. 5d.
per lineal foot.
Granite channel blocks, 12 in. by 6 in., 1s. per lineal foot.
„   „   12 in. by 6 in. (circular), 1s. 3d. per
lineal foot.
„   „   15 in. by 8 in., 1s. 5d. per lineal foot.
„   „   15 in. by 8 in. (circular, $2\frac{1}{2}$ in. to $2\frac{3}{4}$ in.
thick), 1s. 9d. per lineal foot.
Caithness flagging, laid complete, 6s. per square yard.
Coverack concrete slabs, 2 in. thick, laid complete, 6s. 3d. per
square yard.
Limestone flagging, laid complete, 9s. 6d. per square yard.

In Birmingham the average cost per square yard per annum of cleansing, watering and macadamising the carriageways for 1880 was 5d.; the maximum cost for any one street for that year was 4s. 3·35d., whilst the first cost of making was about 2s. 3d. per square yard. In streets of heavy and concentrated traffic, like Bull-street and High-street, the respective costs for macadam, wood and granite were* :—

Material	First cost per square yard.	Average annual maintenance, including first cost and repayment in sixteen years.			Cleansing.		Watering.
	s. d.	s. d.			s. d.		
Macadam	3 0	3 0	per square yard		1 3		·35
Wood	15 0	1 11$\frac{3}{4}$	„	„ „	0 5		·25
Granite	14 0	1 4	„	„ „	0 5		·25

* "Proceedings of the Association of Municipal and County Engineers," vol. vii., p. 81.

A macadamised street in Bristol, 718 yards long, 13 yards wide, cost per annum nearly £1,000 to repair and maintain, or 2s. 1½d. per square yard = £2,425 per mile. It is also on record that Regent-street, when macadamised, cost 3s. 7d. per square yard per annum to maintain.

In Norwich* the cost of constructing syenitic granite macadam, including gravel foundation, is 4s. 6d. per square yard; tarred macadam costs 6s. per square yard, including 2s. for gravel foundation. The tarred macadam requires refacing once in five years, and with this attention has a life of about twenty years in the class of streets where it is suitable.

ROAD-ROLLING.

Some particulars of the history of the evolution of the practice of steam road-rolling have already been given, and need not again be referred to. The use of steam rollers in the consolidation of road metal has become general, both in urban districts and also upon county main roads.

The undoubted advantages of steam road-rolling over the former method of allowing road surfaces to be consolidated by the traffic are briefly as follows:—

(1) Economy combined with efficiency; the roads are better made, and the necessity for such frequent sweeping and scraping is obviated. The saving effected is given as varying from 30 to 50 per cent. The road metal is economised, as a thinner coating can be used, the metal need not be broken so small, and there is less abrasion of the stones, there being only one surface exposed.

(2) The cruelty inflicted upon animals passing over a newly-metalled unrolled road is avoided.

(3) Road-making and repairs can be carried out at any time of the year, and the constant employment of men raking metal into the ruts is avoided.

(4) The steam roller soon shows which is good and which is bad metal for the roads.

---

* "Proceedings of the Association of Municipal and County Engineers," vol. xxii.

(5) Steam-rolled streets are easier for the traffic, are harder, of a more even surface, and have also a better appearance.

(6) The roller is oftentimes found useful for other work.

When thrown open to traffic an unrolled road presents a surface of stones without any mutual cohesion, and every wheel in passing over the loose stones acts somewhat like a plough, pressing down the stones over which it passes and raising up those on each side. This requires the stones to be constantly raked smooth, whereby fresh corners are presented for the next wheels to chip off, and ultimately the surface of the road is uneven, consisting of minute hills and valleys, as it were, which make it far more vulnerable to traffic than when perfectly smooth.

The interspaces in a layer of newly-broken stones occupy, when the stones are first loosely spread upon the road, about one-half the area covered by them, but only one-fourth the area after they have settled into the compactness of an ordinary road surface. This compactness has to be attained by compression or by wear, by a heavy roller, or by the traffic. In the former case the stones are driven into a prepared compressible bed, with their sharp ends downwards and their flat sides uppermost, and form a level, regular and solid pavement of interlocked angular stones; in the latter case they are rolled and kicked about by the traffic until at least one-third of the metalling is destroyed—ground into dust or mud—and removed as refuse by the sweepers; and when, with their angles rounded off, the stones are worn into the worst possible shape for consolidation they are stamped into place to form a road surface, which no mending, filling-up of hollows or other expensive attention will make smooth and durable, and over which the first dry weather of summer will again set the loose stones rolling.

"The main difference between an unrolled and rolled road, at the outset, is that the first contains nearly three times more empty space than the latter. It is clear that a road cannot be hard and strong until these spaces are filled up. Without the use of rolling this can only be done by the particles ground by the traffic off the edges of the stones, by dirt and foul excrementitious matter .... The main causes of the longer duration afforded by the roller are, therefore: (1) That it diminishes the actual wear by the traffic; (2) the interlocking of the stones prevents the injurious action of mud, dirt and moisture; (3) that it allows thinner coatings to be used.

" One of the main advantages attending the rolling of roads by steam-power consists in the diminished proportion of mud or soluble matter which in then incorporated in the structure of the road surface. If the surface of an ordinary road that has not been rolled is broken up and the material washed it is found that as much as half of it is soluble matter—mud, dirt and very fine sand; the stones, having only been thrown loosely upon the road, have lain so long before becoming consolidated by the traffic, and have undergone in the meantime such extensive abrasion, that the proportion of mud, dirt and pulverised material in the metalling is increased to that extent, and the stones are really only stuck together by the mud. This accounts for the fact that although an unrolled macadamised road may, indeed, after long use have a surface that is pretty good and hard in dry weather, and may offer then a very slight resistance to traction, yet it will quickly become soft and muddy when there is any rain. By the employment, however, of a steam roller upon the newly-laid metalling of a macadamised road the stones are rolled in and well bedded at once, and the surface is thus consolidated into a sort of stone felt, capable of resisting most effectually the action of ordinary traffic and containing the smallest quantity of soluble matter to form mud in wet weather."*

The disadvantages attending the use of steam rollers include the following:—

(1) Risk of damage to gas and water mains and services.

---

* " Report on the Economy of Road Maintenance and Horse Draught through Steam Road-rolling, with special reference to the Metropolis," by F. A. Paget, c.e. 1870.

(2) Interference to traffic and the risk of frightening horses whilst the roller is in use.

(3) Nuisance from noise and smoke, though the latter is reduced to a minimum by the use of wood and coke for fuel.

(4) The road metal may be crushed instead of bedded, or the road foundation may be injured, if the roller used is too heavy.

(5) The first cost is a difficulty in small districts which often delays the introduction of a steam roller.

A road-roller should not exceed about 12 tons in weight, or the road metal may be crushed and damage done to gas, water and other pipes, as well as to culverts or cellar arches under the roadway.

The following are the particulars of one of Messrs. Aveling & Porter's modern type 12-ton rollers, well suited to use on county roads :—

Weight, about 12 tons.
Bearing weight on road, 401·2 lb. per square inch.
Length over all, 18 ft. 6 in.
Width of roller surface, 6 ft. 5 in.
Height of top of funnel from ground, 9 ft. 10 in.
Six horse-power nominal.
Driving wheels, 5 ft. 6 in. diameter, 1 ft. 5 in. wide.
Front rollers, 4 ft. diameter, 2 ft. wide.
Boiler made of special brand milled steel.
Plates flanged by hydraulic flanging press.
Large grate area and heating surface.
Gearing made of fine crucible cast steel.
Crank and intermediate shafts carried by patent steel brackets.

Sides of fire-box being intended to form brackets.
Constructed with feed pump and injector for filling.

The fire-box, being extended to form the brackets, allows the gear to be brought closer together, thus reducing the width of the roller, so that whilst narrow streets can be rolled easily the shape of the road is retained.

Messrs. Aveling & Porter, in their pamphlet on "Steam Road-rolling," give the following general description of the manner in which the roller should be used:—

"In the best practice the roadway is excavated, graded and properly formed to a depth of 14 in. from the level of the gutters, with a cross-section conforming to the cross-section of the road when finished, it is then thoroughly and repeatedly rolled with the steam roller, all depressions being carefully filled and rolled before the stone is put on.

"On the bed thus formed and consolidated a layer of stones, 8 in. thick, is set by hand, and rammed or settled to place by sledge hammers, all irregularities of surface being broken off and the interstices wedged with pieces of stone. The intermediate layer of broken stone, of a size not exceeding 3 in. in diameter, is then evenly spread to a depth of 4 in. and

FIG. 31.

thoroughly rolled, and this is followed by rolling in ½ in. of sand. The surface layer of stone, broken to a size not larger than 2 in. diameter and to a form as nearly cubical as possible, is then put on to a depth of 3 in., thoroughly rolled, and followed as before by sand, also rolled. Finally a binding, composed of clean, sharp sand, is then applied, well watered, and most thoroughly rolled with the steam roller, until the surface becomes firm, compact and smooth, the superfluous binding material being swept off and removed."

In "patching" a road by the aid of the steam roller the usual process is, as described by the county surveyor of Notts in a paper on "Steam Rolling,"* to thoroughly water the road surface before applying the new material, and then

---

* "Proceedings of the Association of Municipal and County Engineer," vol. xxii.

"opening the edge of the hole before covering with a patch, removing the fine detritus, applying the material of the gauge required, and covering the edges up with the removed detritus, blending, well watering, and finally well consolidating the whole with a roller."

As to the use of the roller in the repair of roads "when a road becomes so full of holes or so worn as to require coating throughout its entire length and width, it should be hacked completely over and raked into a segmental form in its transverse section to remove irregularities, and so that the road may have a fall from the crown to the channel of not less than 1 in. to 1 yard. It should then be coated with stone broken as nearly cubical as possible and to an uniform gauge. When spread it should be slightly coated with gravel screenings, or the grit sweepings from the roads, which are equally suitable for the purpose when in proper condition. The road should then be watered and rolled, beginning with the road at the channels and ending at the crown of the road, until a smooth surface is obtained, more stones being added to fill up any inequalities that may exist until the whole is consolidated. By constantly sweeping the grit from the sides to the crown of the road as the roller passes over every stone is thoroughly grouted into its bed.*

On the repair of roads the city surveyor of Gloucester observes†: "The road should be thoroughly well lifted and the metalling spread in 3-in. layers evenly, and rolled once or twice before the gravel or other binding material is spread; then spread gravel or sand evenly and well watered with fine distribution until the stone is entirely covered and the sand does not adhere to the roller. Dam up the road channels to prevent water and sand running off into the sewers, and let men scoop up the water and throw it back on the road as it collects in the gutters."

The haunches of the road should always be rolled before the centre, so that when the roller passes over the crown of

---

* "Steam Road-rolling" (Aveling & Porter).
† "The Use of Steam Rollers," by A. W. Parry, Reading.

the road the weight, which will tend to spread the road metal toward the channels, will be resisted by the previously consolidated sides.

"Binding material," consisting of fine gravel, sand, chippings or road drift is essential in road-making with a roller, but should not be placed on the road until the roller has been at work for a short time, as its use is simply to bind the crust or surface of the road. Should too much binding material be used, it will be removed from the joints of the stones by heavy rains and the surface of the road will quickly go to pieces. Where the traffic is allowed to consolidate the newly-laid metal the stones are abraded against each other so as to form sufficient binding material for themselves.

An experiment was tried at New York* in consolidating by rolling of road metal by adhering strictly to the method practised by Macadam of excluding all binding material, but, although the bottom layer of stone could be compacted in this way, it was found impracticable to consolidate the top layer to a sufficiently firm surface to prevent the stones being displaced by the traffic. Increasing the weight of the roller produced the opposite effect to that intended, and the stones became rounded by excessive abrasion. It was thus shown that "broken stones of the ordinary sizes and of the very best quality for wear and durability, with the greatest care and attention to all the necessary conditions of rolling and compression, would not consolidate in the effectual manner required for the surface of a road while entirely isolated from and independent of other substances. The utmost efforts to compress and solidify them while in this condition after a certain limit had been reached were unavailing."

In regard to the rolling of country roads in America, the "Provincial Instructor in Road-making," Ontario, in his report for the year 1896, observes: "When the benefit to be derived from the use of rollers is better understood they will be more generally adopted." The advantages derived from their use are:—

---

* "Roads, Streets and Pavements," by Q. A. Gillmore.

(a) "A good track is immediately obtained, and vehicles at once take the centre of the road.
(b) "A dirt track is not made near the ditch, and by this means the side of the road is not cut up and made so uneven as to interfere with surface drainage.
(c) "Traffic is not inconvenienced in fall by having to drive through loose gravel or crushed stone.
(d) "The gravel or stone is not forced down into the subsoil by the wheels and feet of the horses, is not churned and mixed with the earth, and there is in this way a great saving in the amount of metal.
(e) "There is a great saving in manual labour, and repairs are more easily and effectually made."

Cost of Steam Road-rolling.—The cost of steam rolling per square yard, including all charges, may be taken as being between ½d. and 1d.; but this and the amount of work done in a given time necessarily varies with the system of working and the principle of calculating chargeable expenses. Some surveyors take the daily cost as consisting of wages, coal and oil for the roller alone; others add a percentage on the cost of the machine to cover depreciation and interest; others, again, include the collateral expenses of watching, watering, and sweeping in the sand, &c., sometimes even adding the cost of the hoggin to the account.

Also, the statements of work done in a given time show equal disparity, as so much depends upon the thickness of metal to be rolled, the proportion of sand or hoggin rolled in with it, the quantity of water used in rolling, the gradients of the roads, the degree of consolidation required for the traffic, and the number and gravity of the interruptions met with.

Under efficient management, where the roller is kept continuously at work by having stretches of road always prepared in advance of it, the number of square yards rolled in a day by Messrs. Aveling & Porter's rollers is given at from 1,000 to 2,000, according to the weight of the machine and to the influence of the other conditions on the work. If the items constituting the cost of rolling be restricted to the wages of the driver, the bills for coke and oil, and a sum of 10s. to 15s. to cover interest, wear and tear, the cost of rolling the above number of yards will average 22s. to 25s.— *i.e.*, from 3 to 8 square yards can be rolled and thoroughly consolidated for 1d.

In Edinburgh 10,000 tons of metal, covering an area of 100,000 super. yards of roadway, were consolidated at a cost, including all expenses, of ½d. per square yard. Two thousand five hundred super. yards of 3-in. or 4-in. metal have been consolidated in a day, but the mean surface rolled has been stated as 1,000 yards for town roads and 2,000 yards for country roads where there is no interruption to the working of the roller.

In Bournemouth 9,666 super. yards were rolled in 116 days, at a cost varying from ¼d. to ½d. per yard super.; 4,800 super. yards were also picked up at a cost of from 1d. to 2d. per yard super.

The cost per day of steam roller in Bristol was reported as follows: Driver, 5s. 6d.; man with flag, 3s.; coke and coal, 2s.; oil and sundries, 2s.; interest, depreciation and repairs, 6s. 1d.; hire of water-cart, 8s.; total, £1 6s. 7d. An area of 2,043 yards super. of roadway was consolidated at a cost of rather more than ½d. per yard super. Another example of 4,145 yards super. cost ·56d.

On the question of cost the following details, given by Mr. A. Greenwell, A.M.I.C.E. (Frome), in a paper on "Steam Road-rolling,"* will be of interest: "From careful experiments with blue limestone it has been found that to obtain consolidation with the usual coating of two stones in thickness (each cubic yard broken to 2½ in. gauge, and made to cover about 1·7 super. yard of road) the steam roller must traverse a patch equal to its own width about thirty-five times. From this it appears that a cubic yard of broken stone requires 1½ ton-miles of rolling to produce consolidation.

"With regard to the use of binding material, the blue limestone in this district (Frome) cannot be consolidated without

---

* "Proceedings of the Association of Municipal and County Engineers," vol. xx.

it. The best results appear to be obtained by the use of well-weathered road scrapings, spread over the surface, when consolidation has been nearly effected, in the proportion of one of road scrapings to twenty of broken stone. Careful observation has shown that at least 25 per cent. of the material is saved by steam road-rolling.

"The cost of steam road-rolling a patch of mountain limestone about two stones in thickness (1 cubic yard broken to 2½ in. gauge, to cover 17½ superficial yards), 1 mile in length and 21 ft. wide, is found to be £31 1s. 7d.

"The cost of this coating is £189 11s. 8d., and under ordinary circumstances, when consolidated by the traffic alone, will last seven years; making the cost per mile per annum £27 1s. 8d. Taking the saving in materials resulting from the use of the steam road-roller at 25 per cent., the additional cost (for steam road-rolling) per mile per annum would be £3 6s. 7d., but the saving in material would be £6 15s. 5d.; and, allowing 16s. 8d. for after-raking, the nett profit resulting from the use of the steam road-roller is £4 5s. 6d., or 15 per cent."

TOWN ROADS.

When it is found that the traffic in a road has so increased as to render it more economical and more advantageous to the public convenience, as well as to that of the residents in the road, to provide it with some harder and more durable surface than macadam, the roadway is then usually pitched with stone, paved with either wood, asphalte, brick or other suitable material.

As in the case of macadamised roads, a good and properly constructed *foundation* is absolutely essential in order to carry the weight of the traffic, the surface material of the roadway being regarded more in the light of a veneer to withstand the *wear* of the traffic and to preserve the foundation underneath.

The best foundation for paved roadways is that composed of cement concrete. This is prepared in various proportions upon "bankers," or boarded platforms, as follows: Four parts of broken stone or brick are mixed with two parts of clean sharp sand and one part of Portland cement. After being turned over twice in a dry state water is added from a watering-pot fitted with a rose, and the mixture again turned over twice in its wet state. Being thoroughly mixed, it is at once placed in the excavation prepared for the foundation, consolidated by ramming, brought to the proper level, and the surface finished with a shovel. Wood pegs are usually inserted in the roadway surface to guide the workmen in attaining the required contour and heights. The concrete should be in position within a quarter of an hour of the time of mixing.

In Liverpool another method is adopted for laying in this foundation; it is that of the *béton* horizontal wall or slightly cambered arch. Mr. H. Percy Boulnois, formerly city engineer, Liverpood, has described this method as follows*:—

" The ground having been prepared in the usual way, and the channel and kerb stones fixed in position, a stratum of stones (which should by preference be of a non-absorbent character), broken so as to pass all ways though a 3-in. ring, is spread evenly over the surface of the ground, and upon this is placed a layer of cement mortar, mixed, in the proportions of one of Portland cement to six of fine, sharp, clean gravel, in the method to be described hereafter. Upon this layer of mortar is placed another layer of broken stone —the whole of the stones in each layer to be thoroughly watered while the work is proceeding—and this stone is forced into the interstices of the first layer by the use of flat beaters of wrought iron weighing 16 lb. each, shaped like square shovels, with handles at an angle of 33°.

" This process is repeated until the proper level and contour is reached, and the surface is finished off parallel to the exact curvature of the carriageway. The foundation thus prepared is left until the concrete is sufficiently set or hardened to receive the pavement, which, if possible, should

---

* " The Construction of Carriageways and Footways," by H. Percy Boulnois, c.e. (Biggs & Co.).

not be less than ten days, although this period may be shortened where the exigencies of the traffic render it imperative by strengthening the proportion of cement to the gravel, care to be taken in all cases to periodically water the surface of the concrete to assist the ultimate hardening, and in very hot weather it is advisable to cover the surface of the concrete with old cement bags thoroughly saturated with water.

"The proportions of broken stone, gravel and cement used in such a concrete are as follows:—

"*Before Mixing.*—Broken stone, eight parts; gravel, six parts; cement, one part.

"*After Mixing.*—Broken stones and gravel, mixed together, eleven parts; cement, one part; three parts of gravel having been expended in filling the interstices of the stones."

Another description of foundation is that known as "bituminous concrete"; it is made as follows:—

The ground is excavated to the required depth and contour, and broken stone as used for macadam is laid in for a depth of from 6 in. to 9 in., and levelled and rolled. A boiling mixture of pitch and tar, or creosote oil, is next poured over the surface to fill all interstices, and a thin layer of stone (broken small) is then spread and well consolidated by rolling. The cost per square yard of bituminous concrete foundation in Liverpool, 6 in. deep, including all charges, is about 3s. 6d.

STONE PAVEMENTS.

A pavement of granite setts, laid upon a good cement concrete foundation, makes the most durable carriageway that can be constructed, and is particularly suited to streets of heavy traffic. It can be used upon all ordinary gradients, is suited to all classes of traffic, and affords ease of traction and a very fair foothold to horses. Also, it is easily cleansed, and creates the minimum of dust and mud.

The objections to this class of pavement are that it becomes greasy and slippery under certain atmospheric conditions, and that the incessant noise from the traffic in any busy thoroughfare so paved is a great inconvenience to tradesmen and others. It is also considered that the jar upon the legs and hoofs of horses is injurious to them.

Granite setts of large dimensions (6 in. to 8 in. wide by 10 in. to 20 in. long by 9 in. deep) were at first employed, but subsequent experience has shown that narrow setts, about 3 in. in width, are much to be preferred and afford a much better foothold to horses. The noise nuisance may also be lessened by running the joints with asphaltic composition instead of ordinary grouting.

The following useful table, given by Mr. H. Percy Boulnois,* gives the sizes of pitchers now in general use, and also shows the number of square yards that 1 ton in weight of the different sizes of setts, cubes and blocks will cover.†

Stone.	Depth × Width × Length.	Area in square yards which 1 ton will pave.
Setts	5 in. × 3 in. × 5 to 7 in.	4·5 square yards.
,,	5¼ in. × 3¼ in. × ,,	4·3 ,,
,,	6¼ in. × 3¼ in. × ,,	3·6 ,,
,,	7¼ in. × 3¼ in. × ,,	3·1 ,,
Cubes	3¼ in. × 3¼ in. × 3¼ in.	6·7 ,,
,,	3½ in. × 3½ in. × 3½ in.	6·2 ,,
,,	3¾ in. × 3¾ in. × 3¾ in.	5·8 ,,
,,	4 in. × 4 in. × 4 in.	5·4 ,,
Blocks	4 in. × 4 in. × 6 in.	3·6 ,,
,,	4 in. × 3 in. × 3 in.	5·4 ,,
,,	5 in. × 3 in. × 3 in.	4·4 ,,
,,	6 in. × 3 in. × 3 in.	3·7 ,,
,,	6¼ in. × 3¼ in. × 3¼ in.	3·25 ,,

The suitable *width* of a stone is to some extent regulated by the size of the horses' hoof, about 4 in. being the maximum. A *depth* of 7 in. is found sufficient for stability under any class of traffic, and the *length* should be such as to properly break joint with the adjoining stones. If too long the stones are apt to tilt and work loose. The stones should be well and uniformly dressed, so as to make as close a joint

---

* "The Municipal and Sanitary Engineers' Hand-Book" (Spon).
† This, of course, will vary with the specific gravity of the stone.

as possible, and should not be allowed to vary but very
slightly from the specified sizes. The finer the joints the less
will be the noise produced by traffic, and the wear will also
be reduced to a minimum. In good work the setts are laid
upon a cushion bed of fine sand ½ in. in thickness, so as to
provide an elastic bed, and to convey the pressure equally to
the foundation.

In regard to the cleaning of old setts for re-use, Mr.
H. Percy Boulnois, when city engineer of Liverpool, found
that the expense of cleaning off the grouting of bitumen, or
cement, by hand with the sett-cleaner's hammer cost 10½d.
per square yard, as compared with 4·07d. per square yard
when the old setts were boiled in a pitch boiler with creosote
oil heated to about 266 deg. Fahr.

Local circumstances often determine the class of stone to
be used as paving material, but care should be taken that
the stone selected should be hard and durable, and not apt to
wear slippery, or brittle. The hard igneous and metamorphic
rocks are the best for pavement setts where the traffic is
heavy, but where comparatively light millstone grit and other
hard sedimentary rocks are used. Carnarvonshire syenite is
one of the most durable materials that can be used for heavy
traffic, and most of the granites are also largely used. Of
the latter, the Balbeattie granite is considered the best.

A pitched roadway, consisting of 7 in. by 3 in. Norway
granite setts, laid on 6 in. of Portland cement concrete, costs
about 17s. per superficial yard in the neighbourhood of
London.

Penmaenmawr stone has been much used in the North,
but was discontinued in London owing to its slipperiness. It
is a hard and durable stone, but wears smooth and is noisy.

Aberdeen blue granite is preferred in London to either
Guernsey or Mountsorrel, as it retains a rougher surface,
but wears faster. Colonel Haywood's observations showed
that in London the wear of Aberdeen granite pavements was
from ·14 in. to ·23 in. per year, whilst the wear of Penmaenmawr and Carnarvonshire setts in the city of Liverpool, under
heavy traffic, is said to seldom exceed ·02 in. per annum.

The cross-section of a paved street should have a rise to the crown of one-sixtieth the width of carriageway.

I am indebted to the courtesy of Mr. John A. Brodie, M.I.C.E., city engineer, Liverpool, for the following particulars of standard specifications for paving of streets in that city, and which will serve as excellent examples of how this class of work should be carried out in districts of this character.

First-Class Specification.

Excavate or fill in the ground, as the case may be, to the requisite level, and remove all surplus material. Properly form and trim off the surface and thoroughly consolidate same, and then lay a foundation of not less than 6 in. of Portland cement concrete, corporation standard. The paving shall consist of granite or syenite setts, $3\frac{1}{4}$ in. wide by $6\frac{1}{4}$ in. deep, from North Wales or other approved quarries, laid in regular, straight and properly-bonded courses, with close joints, and to be evenly bedded on a layer of fine gravel $\frac{1}{2}$ in. in thickness. After the paving is laid the joints shall be filled with hard, clean, dry shingle; the setts shall then be thoroughly rammed, and additional shingle added until the joints are perfectly full. The joints shall then be carefully grouted until completely filled with a hot composition consisting of coal pitch and creosote oil, and finally the paving is to be covered with $\frac{1}{2}$ in. of sharp gravel.

The crossings shall consist of three rows of 16 in. by 8 in. granite crossing stones, and the remaining space shall be paved on each side of the crossing stones, to the full width of the footway, in a similar manner to the carriageway. The crossing stones shall be of granite of a quality to be approved by the city engineer, dressed perfectly true, and out of winding on the face; the sides and joints to be perfectly square and accurately dressed throughout their entire depth; the stones to be bedded on cement concrete, the joints to be filled with shingle and grouted in a similar manner to the paving. A triangular groove, 1 in. wide by $\frac{3}{4}$ in. deep, to be formed along the upper surface of each stone. No stone to be less than 3 ft. in length.

The footways shall be paved with Lancashire ("Best Barns") or Yorkshire flags of the best quality, not less than 3 in. thick. No flag to measure less than 2 ft. in width nor to be of less area than 6 ft.; to be solid, free from laminations, the upper surface to be true and free from windings or hollows; the joints to be squared the whole thickness. The flags to be laid on a bed of fine gravel, with close, neat joints flushed in mortar, and in uniform courses breaking bond. The joints to be dressed after laying, where necessary.

The channel stones to be of granite or syenite, of a quality to be approved by the city engineer, and to be not less than 3 ft. in length. The upper surface, if not self-faced and perpectly true, must be accurately worked out of winding, the bed even and parallel to the face, the sides and ends truly square; the stones to be bedded on cement concrete, and the joints to be filled with clean shingle and grouted in a similar manner to the paving.

The curb stones to be granite or syenite, straight or circular as required, 6 in. thick at top, 7 in. thick at 5 in. below, and not less than that thickness for the remainder of the depth; to be not less than 12 in. deep, nor less than 3 ft. in length; to be carefully dressed on top, 8 in. down the face and 3 in. down the back; the remainder of each stone to be hammer-dressed; the heading joints to be neatly and accurately squared throughout the entire depth.

SECOND-CLASS SPECIFICATION.

Excavate or fill in the ground, as the case may be, to the requisite level, and remove all surplus material. Properly form and trim off the surface and thoroughly consolidate the same, and then lay a foundation of (a) not less than 6 in. of Portland cement concrete, corporation standard, or (b) not less than 6 in. of bituminous concrete, consisting of clean and angular broken stone, grouted with a hot composition consisting of coal pitch and creosote oil, covered with chippings and thoroughly consolidated by rolling with a roller of sufficient weight. The paving shall consist of granite or syenite

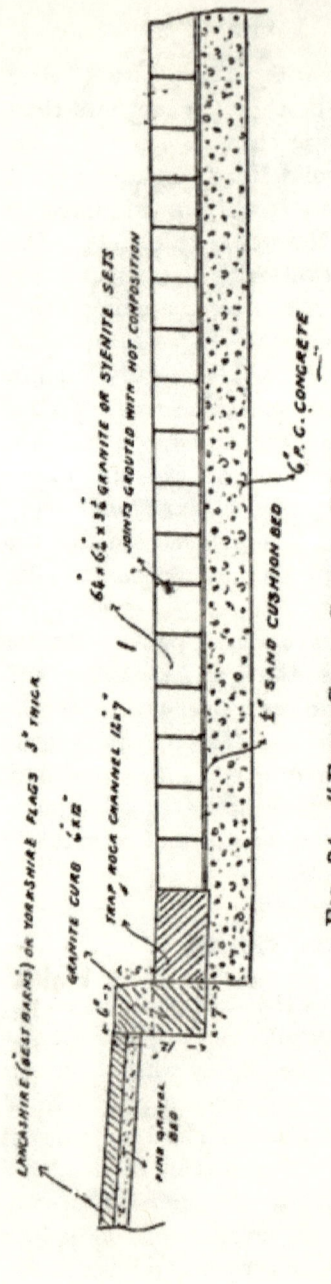

Fig. 34.—"First-Class Streets," Liverpool.

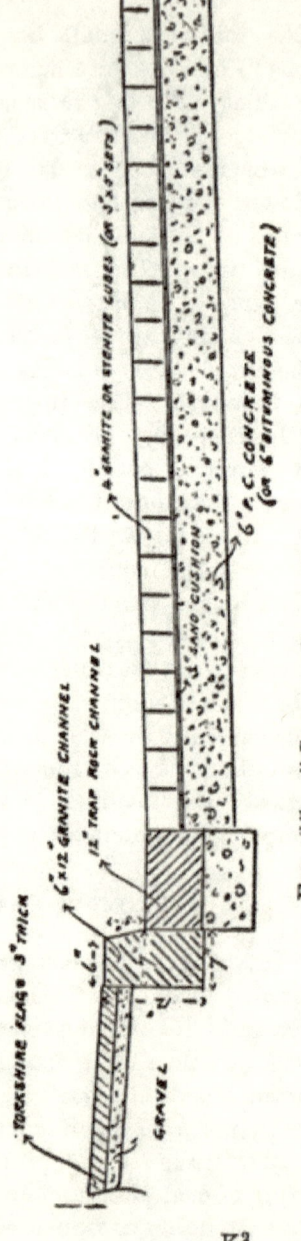

Fig. 35.—"Second-Class Streets," Liverpool.

setts 3 in. wide by 5 in. deep, or of granite or syenite, 4 in. by 4 in. cubes, from North Wales or other approved quarries, laid in regular, straight and properly-bonded courses, with close joints, and to be evenly bedded on a layer of fine gravel, ½ in. in thickness. After the paving is laid the joints shall be filled with clean, hard, dry shingle; the setts shall then be thoroughly rammed, and additional shingle added until the joints are perfectly full. The joints shall then be carefully grouted until completely filled with a hot composition consisting of coal pitch and creosote oil, and, finally, the paving shall be covered with ½ in. of sharp gravel.

The crossings, footways, channels and curbs shall be the same as specified for first-class streets.

Third-Class Specification.

Excavate or fill in the ground, as the case may be, to the requisite level, and remove all surplus material. Properly form and trim off the surface, and thoroughly consolidate the same, and then lay a foundation of hand-pitched rock, 10 in. in depth, set on edge in the manner of a rough pavement. Over this a coating of gravel is to be laid of sufficient thickness to fill in the interstices and to form a smooth surface to the foundation, which must be thoroughly consolidated by rolling with a steam roller before the paving is laid. The paving shall consist of 4 in. by 4 in. granite or syenite cubes, from North Wales or other approved quarries, laid in regular, straight and properly-bonded courses, with close joints, and to be evenly bedded on a layer of fine gravel, ½ in. in thickness. After the paving is laid the joints shall be filled with clean, hard, dry shingle; the setts shall then be thoroughly rammed, and additional shingle added until the joints are perfectly full. The joints shall then be carefully grouted, until completely filled up, with a hot composition consisting of coal pitch and creosote oil, and, finally, the paving shall be covered with ½-in. of sharp gravel.

The crossings, footways, channels and curbs shall be the same as specified for first-class streets.

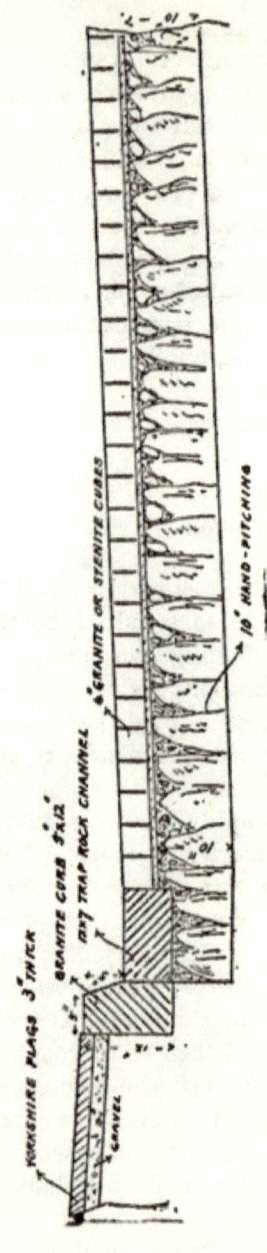

Fig. 36.—"Third-Class Streets," Liverpool.

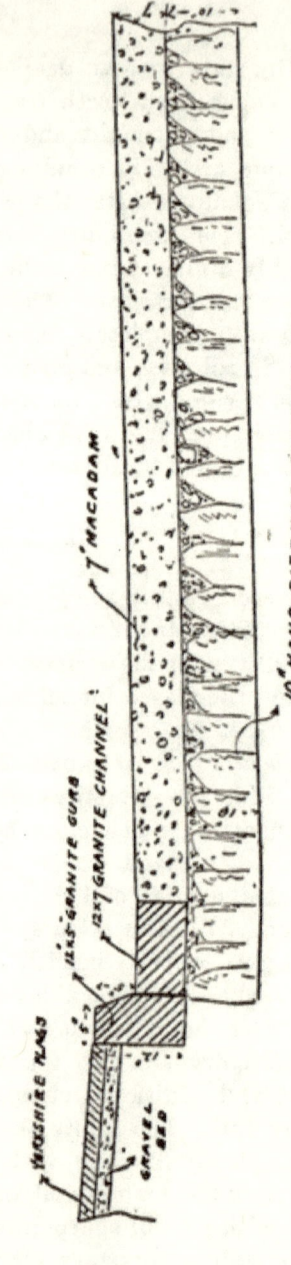

Fig. 37.—"Fourth-Class Streets," Liverpool.

Fourth-Class Specification.

After excavating the area of carriageway to a depth of 17 in. below the finished surface line, hand pitch the same with hard rock set on edge with the broadest sides downwards, 10 in. deep, break off the irregular corners of the stones, and fill in the interstices of the pitching with these fragments as well as with similar rock broken small. Consolidate the foundation thus formed by passing over it a steam roller until the whole is firm and compact. When this is done spread over the surface evenly to a depth of 7 in. broken screened macadam free from slaty or flat fragments. This macadam to be durable granite or trap rock from the quarries of North Wales or from other approved quarries having similar class of rock. A layer of $2\frac{1}{2}$-in. gauge stones to a depth of $3\frac{1}{2}$ in. to be first spread and rolled until solid with a steam roller, then a layer of 2-in. gauge stones to the finished level consolidated in a similar manner. Where binding is necessary, it should be sparingly used and must consist of granite chippings, preferably of the same rock as the macadam.

Lay channel stones over the hand pitching on each side of the carriageway of granite or syenite, of a quality approved by the city engineer, in lengths not less than 3 ft., in depth not less than 7 in., in width 12 in. The upper surface, if not natural faced and perfectly true, must be accurately worked out of winding, the bed as far as practicable parallel to the face, the sides and ends truly square, and the joints filled with clean shingle grouted with melted pitch and creosote oil.

Lay granite or syenite curb stones of a similar quality to that specified for channels, straight or circular as required, 5 in. thick at top, 6 in. thick at 5 in. below, and not less than that thickness for the remainder of the depth; to be not less than 12 in. deep, nor less than 3 ft. in length; to be carefully dressed on top, 8 in. down the face and 3 in. down the back; the remainder to be hammer dressed, the heading joints to be squared throughout the entire depth.

Pave the footways with "Best Barns" Lancashire flags, or Yorshire flags of the best quality, not less than 3 in. thick,

2 ft. in width, nor less than 6 ft. superficial area in each flag, which must be solid, free from laminations, windings, or hollows on the surface; the joints to be squared the full thickness; the flags to be truly laid on a bed of fine gravel, with close joints flushed with hydraulic lime mortar and in uniform courses, breaking bond; the surface to be flogged off after laying where necessary, but great care to be exercised in bedding the flags, so as to prevent the necessity for the after flogging. Natural asphalte upon concrete, granolithic or other artificial paving may be laid in substitution of flags at the discretion of the city engineer, and in accordance with his directions in each case.

Lay crossings, where required, consisting of three rows of 16-in. by 18-in. granite of a quality approved by the city engineer, and in lengths not less than 3 ft., dressed true and out of winding on the face, the sides and joints square, and accurately dressed throughout their entire depth; the joints to be filled with clean shingle, and grouted with pitch and creosote oil; a V grove, 1 in. wide and ¾ in. deep, to be cut along the surface of each stone.

Fifth-Class Specification.

(Specification for passages or back streets 9 ft. wide and upwards.)

After excavating the area of carriageway to a depth of 17 in. below the finished surface line, hand pitch the same with hard rock set on edge, with the broadest sides downwards, 10 in. deep, break off the irregular corners of the stones, and fill in the interstices of the pitching with these fragments as well as with similar rock broken small. Consolidate the foundation thus formed by passing over it a steam roller until the whole is firm and compact.

The paving shall consist of setts of Haslingden grit stone of the best quality, uniformly gauged to a depth of 6 in., laid in courses 6 in. wide, with a fall towards the centre of 1 in 48. Two courses to be laid longitudinally, to form a channel. After the paving is laid the joints shall be filled with hard, clean, dry shingle; the setts shall then be thoroughly rammed,

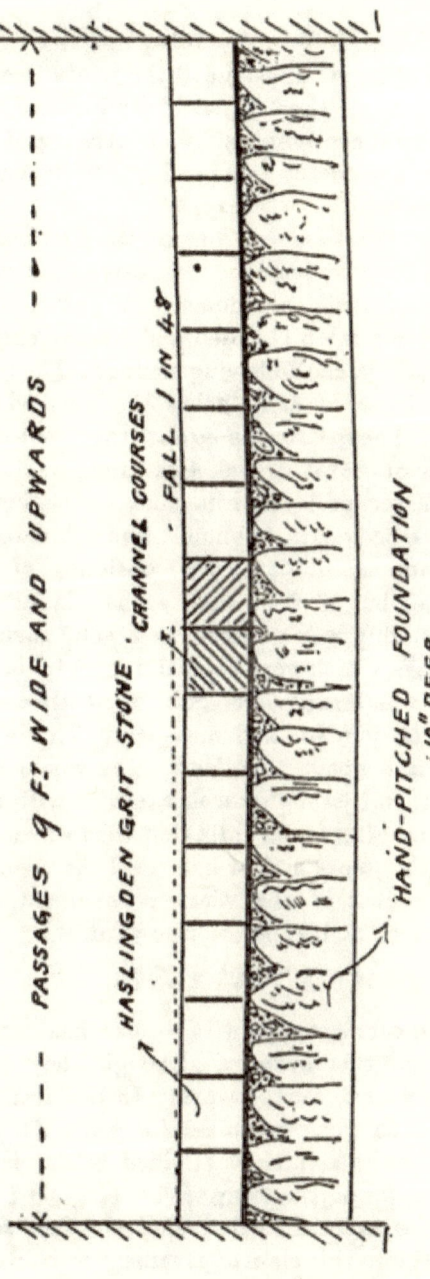

Fig. 38.—"Fifth-Class Streets," Liverpool.
(Passages 9 ft. wide and upwards.)

and additional shingle added until the joints are perfectly full. The joints shall then be carefully grouted until completely filled with a hot composition consisting of coal pitch and creosote oil, and finally the paving is to be covered with ½ in. of sharp gravel.

The "*Euston*" *Pavement* was one of the first good granite pavements laid down in London. It derived its name from the fact of having been laid down (1843) at the departure side of the Euston station (London and North-Western Railway). It was laid in the following manner: The ground was excavated to a depth of 16 in. below the finished surface of the pavement and shaped to the proper road contour. Upon this bed a layer of coarse gravel, 4 in. thick, was spread and well rammed. A second layer, 4 in. thick, consisting of gravel mixed with chalk or hoggin, to blind it, and again well rammed was next spread. A similar layer consisting of the same kind of materials but of finer quality was placed upon this, and again thoroughly well rammed to a solid surface for receiving the stones, which were bedded upon 1 in. of fine sand. The stone used was Mountsorrel granite, neatly dressed, and squared and well jointed, and measuring 3 in. in width by 4 in. in length and about 4 in. deep. The whole surface of pavement was then thoroughly well rammed with a rammer, 55 lb. weight, and afterwards sprinkled with screened gravel, which entered the joints and so increased the rigidity of the pavement. The cost of the Euston pavement, including foundation, is given at 12s. per square yard.

BRICK PAVEMENTS.

The paving of carriageways with bricks has received but little attention in this country, although they have been somewhat largely used for footways. In Holland the use of brick for this purpose has prevailed for some 150 years, and in America and other countries "vitrified brick" is also very largely employed for carriageway traffic.

In England the experience derived from the use of brick upon footways shows this class of pavement to be particularly apt to wear unevenly, owing to the varying quality of the bricks. The want of attention of English municipal engineers to the use of this material has been pointed out by Mr. J. T. Eayrs, M.I.C.E., in a very interesting paper, entitled "Brick Paving for Carriageways,"* in which he says in America "It is claimed that neither granite, asphalte nor wood can offer so many advantages as vitrified brick as a paving material, and that, if properly laid, it is as noiseless as any other kind of pavement: the surface is smooth without being slippery; it offers a minimum amount of resistance to the passage of traffic, and inflicts a minimum amount of wear and tear on horses and vehicles; it is practically impervious, and therefore perfectly sanitary; is easily cleansed, and requires less scavenging than any other paving; it can be washed without injury or becoming slippery when wet; is readily taken up and relaid; reasonable in first cost and maintenance; and has a life which compares favourably with other materials, such as asphalte, wood, &c."

The city engineer of Chicago says, "Comparing brick pavements with asphalte, I believe that for general use shale brick, properly burned and of the right size, properly laid on a hydraulic cement concrete foundation, is superior to asphalte, as to its first cost, facilities and cheapness for repairs, its sanitary qualities, its ease upon horses, and last, but not least, its durability. . . . Brick can be laid on a grade where it is out of the question to lay asphalt. Brick can be washed continually without injury. . . . Brick when wet is not more slippery than when dry; asphalte is always dangerous, and if wet is more slippery than when dry, and when a horse is down on asphalte it is with difficulty that he regains his feet. Brick is not more noisy than asphalte, and as it can be continually sprinkled it is far less dusty."

In selecting a suitable brick for paving purposes the points to which the attention should be directed are—the nature of the clay from which they are manufactured, the uniformity of shape and size, absorption, specific gravity, transverse and crushing strength and abrasion.

* "Proceedings of the Association of Municipal and County Engineers," vol. xxiii.

In regard to the quality of the clay, Mr. Eayrs, in the paper above referred to, points out, " The composition of this varies very considerably in different districts, but most bricks are made from a hydrated silica of alumina, generally containing traces of magnesia, iron, lime and potash. Lime in excess is very injurious to paving brick, as it is changed to caustic lime in burning, and a small amount of moisture will cause it to slake and disintegrate the brick. A small amount of magnesia aids in producing vitrification. Iron is not injurious, but in a brick to resist high temperature an excessive quantity would be fatal. Alumina gives elasticity, and renders the material tough and binding. The alkalies act as a flux in chemical combination with the silica and alumina. One writer states that from an analysis of a large number of clays, most of them carboniferous, he finds, after averaging one with another, that alumina, silica and water make up about 85 per cent. of the material, leaving 15 per cent. for lime, potash, soda, iron and other impurities. Clays with a low percentage of these impurities, and more especially of potash, soda and iron, are fireclays. Alumina and silica are practically infusible alone, but the presence of even 3 per cent. of potash or soda renders the whole mass easily fusible. The presence of these so-called impurities is absolutely essential where vitrification is desired. The ease with which a clay will vitrify depends largely upon the percentage of the respective impurities which act as fluxes. A clay may have so large a percentage of fluxing material as to make it too-easily fusible."

In America the sizes of paving bricks vary between $7\frac{1}{2}$ in. by $3\frac{1}{2}$ in. by 2 in. and $9\frac{1}{4}$ in. by $4\frac{1}{2}$ in. by $3\frac{1}{4}$ in., but the size mostly used is 9 in. by 4 in. by 3 in. They are made both with square edges and rounded nosings and with rounded tops. The square edges are mostly used, make better joints, and are more easily scavenged.

Paving bricks should not absorb more than from $1\frac{1}{2}$ to $3\frac{1}{2}$ per cent. of moisture in twenty-four hours. The degree of absorption is an important factor in determining the life of the pavement, as the succession of wet and dry weather, and of frost and thaw, tend to disintegrate the brick. Where the specific gravity of a brick is specified, it is required in American specifications to be from 2·00 to 2·30. A transverse strength of about 2,000 lb. is also required, and a crushing resistance of from 10,000 to 12,000 lb. per square inch.

The greatest amount of wear on a road surface is caused by the pounding produced by horses' feet, so that paving bricks which will withstand an abrasion test will have the longest life. For this test the bricks are usually placed in a "tumbler" or "rattler," either with or without other materials. The tumbler, consisting of an iron barrel mounted on trunnions, is made to revolve at a certain speed and for a certain time, and in so doing rattles the bricks together with iron castings, or other materials placed inside with the bricks, so as to abrade the surfaces. The bricks under test are weighed before being placed in the tumbler, and again after rattling for a specified time, and the loss is then noted.

In laying brick pavements in America the only foundation adopted in many instances consists of either macadam, broken ballast, slag or gravel rolled down. In some cases the bricks have been laid without foundation, except the rolling of the natural ground and spreading a bed of sand. The necessity of a stronger foundation, however, is now being recognised, and the larger cities are adopting a Portland cement concrete foundation with a cushion, 1 in. to 2 in. in thickness, of clean, dry sand to bed the bricks upon. About seven days are allowed for the concrete to set before the bricks are laid.

The bricks are laid in straight courses at right angles to the curb, or diagonally at an angle of 60°. They are usually laid close together, without leaving spaces for joints, the irregularity in shape of the bricks being of itself sufficient for this. In some instances the bricks are even forced up towards the curb with the aid of a lever, and then wedged up with a closer; but this practice seems likely to cause the paving to lift and arch, or to displace the curbs in the event of expansion.

The bricks are next either hand-rammed or rolled with a steam or horse roller drawn by men. In ramming, a plank of 1 in. or more in thickness is frequently placed between the bricks and the rammer.

The pavement is grouted with either cement or pitch and sand. The sand used is clean, sharp river sand. In pitch grouting a semi-elastic watertight joint is aimed at, such as will not crack in winter or "spew" up in the hot season. The cement grout is mixed with sand in the proportion of from one to one and one to one and a half. The pitch or cement grouting is applied by means of a filler or funnel-shaped instrument, so that the grout is not allowed to run over the pavement, but simply to fill the joints. The curvature of cross-section of the pavement varies with the gradient, but is the same as that adopted in wood paving.

The pavement is finally covered with clean, coarse sand ($\frac{1}{2}$ in. to 1 in. in thickness), and allowed to so remain, with traffic passing over it, for about a fortnight, when it is cleared off.

This pavement is usually laid under contract, including maintenance after completion for periods from one up to five years. The life of brick paving in America is given as ranging from fifteen to twenty years under ordinary conditions, and its use for carriageways appears to be rapidly increasing.

The provincial instructor in road-making to the Ontario Department of Agriculture, in his report on "Road and Street Improvement in Ontario" for the year 1896, contains some instructive observations on the question of the use of vitrified brick for street pavements in that province. He says:—

"The only competitor of asphalte is vitrified brick. This pavement has come in a most timely way to take the place of the decayed cedar-block pavements, which are disgracing the streets of so many towns and cities. Vitrified brick is becoming popular, and presents features which tend to cause it to become more so. It offers a better foothold for horses than does asphalte. The surface is not so smooth, and in consequence radiates less heat and light, is quite as sanitary, with less liability to become dusty. Among bicyclists it is much more popular than the asphalte. If the joints are filled with suitable cement brick pavement is but little more noisy than asphalte.

"The majority of failures which have occurred with brick have been traced to defects which the material or better construction could have obviated. Its ease of construction and repair offers a great advantage over asphalte, ordinary labourers being easily taught to do the work. Few repairs are needed if good brick is used, and in the first cost as well as in maintenance brick should be, and generally is, cheaper than asphalte.

"Although brick is one of the oldest paving materials, it has been used on this continent for only about a quarter of a century; and only within the last ten years has it attracted widespread attention. In the United States it has been used very extensively, but in Ontario experience with the modern vitrified brick is very limited.

"Of its success as a paving material little remains in question. Brick pavements have been in existence in the United States for eighteen years, remaining in good condition. It was feared that the climate of northern countries, with severe frosts and rapidly-alternating conditions of moisture and temperature, would be unfavourable to its use, but the experience of various northern cities shows that vitrified brick of a good quality is a most valuable addition to our list of paving materials.

"The best vitrified brick is made of shale or clay, or a mixture of the two. It is not "vitrified," as the name indicates, but is raised by intense heat just to the point of fusion. More than this fuses or melts the clay, permits it to run together, and the product is then glassy or vitrified, and brittle in consequence. The process of cooling must be very gradual. A brick if too rapidly cooled or "annealed" will be brittle, but with a thoroughly pulverised and well-mixed shale, brought to the proper temperature and then slowly annealed, the resultant brick should be sufficiently hard and tough to scratch steel. . . . The chief tests used are—that of absorption, representing the probable effect of atmospheric action, the rattler test, showing the effect of impact and abrasion as found in the chipping of horses' hoofs and the grinding of wheels, the transverse strength, showing the power to resist the breaking strain of heavy loads. Other tests are sometimes used, such as to determine the crushing strength, and the depth to which oil will penetrate the surface, &c.

"Where in towns the facilities for performing these experiments are not available, and only short sections of pavement are to be constructed, the experience of cities will generally afford a safe guide in choosing between different makes of brick. A further safeguard may be had by requiring the contractor or manufacturer to maintain the pavement up to a certain standard for at least five years.

"There is a tendency to endeavour to reduce the cost of brick pavements by the use of a weak foundation or no foundation at all. Brick pavements laid on gravel and sand have been successful, but this has been the case only when the subsoil has been of such a kind as to be very porous, easily drained and naturally firm. The experiment in Ontario, in view of Fall and Spring conditions, with alternate freezing and thawing, is a very dangerous one, a lesson which has been strongly impressed by the experience with cedar blocks. While a brick pavement may give satisfaction for a few years on sand and gravel foundation, there is every probability that the brick will settle irregularly, and rendered thereby more susceptible to wear and strain, the bricks will be broken and the life of the pavement very much shortened.

"A concrete foundation should almost invariably be employed. A 4-in. layer will, where a brick surface is suitable, be sufficiently durable. This forms a stiff monolith base, which distributes the weight of the traffic. There cannot be irregular settlements of brick, as is the tendency with yielding materials, such as sand and gravel. It also prevents water percolating beneath the road—not a very important feature in the South, but in freezing climates a matter of considerable importance.

"Between the bed of cement and the surface covering of brick a thin cushion of sand is necessary. By this means the brick can be laid evenly, a certain amount of spring is obtained, which lessens the effect of blows on the brick, and it overcomes the rumbling noise otherwise created. It is a common practice to merely fill the joints of the brick with sand. While this is not at all objectionable, by the use of a cement composed of pitch and sand the pavement becomes less noisy, absorbs less street filth, and corners and edges of the brick are strengthened."

McDougall's Patent "Combination Set Pavement."—This consists of a highly vitrified blue paving brick, 10 in. by $4\frac{1}{2}$ in. by 5 in., containing recesses into which square wooden plugs, 2 in. long by 1 in. square, are driven, thus giving it the character of a wood pavement rather than brick. The plugs, which are previously creosoted, are intended to remain always about $\frac{1}{16}$ in. above the surface of the brick. The advantages claimed for this type of paving are: That it is not slippery and affords a good foothold for horses, that it is durable, non-absorbent, wears evenly, is cheap, and readily handled and laid. It is laid on a concrete foundation with close joints, and grouted with hot pitch or a bituminous mixture. This pavement has been used at Cheltenham, Oxford, Preston, Manchester, Bootle and elsewhere.

At an annual meeting of municipal engineers, held in London in July, 1897, Mr. J. Hall, borough surveyor, Cheltenham, said: "I have laid several crossings with McDougall's patent bricks, and after two years' experience a motion has been adopted by the council to the effect that no other material is to be used anywhere for street crossings. The first crossing has been under observation by the police, and no record of any horse falling upon it has been made during that time, though previously it was a common experience to have two or three horses down upon a granite sett crossing in the same place in a week. The cost is about the same as granite."

WOOD PAVEMENTS.

The use of wood as a carriageway pavement has very largely increased of recent years, due, most probably, to its being absolutely the least noisy material to hand, and to its being free from the drawback of that degree of slipperiness attributed to asphalte. Also, with the exception of asphalte, it gives the minimum of traction.

Reference to the early introduction and use of wood has already been made, since which time very marked improvements have been made in the methods of laying, as a result of the vast amount of experience which has been gained in the employment of this class of pavement in the metropolis and numerous other places through the country. A fresh impetus to the use of this material has of late years been afforded by the introduction of the West Australian and New South Wales hard woods, which are now being somewhat extensively adopted both in London and provincial towns.

For streets of medium traffic wood is undoubtedly an excellent pavement, but is unsuited for streets having a heavy mercantile traffic of, in some cases, from 4 to 6 tons per wheel, such as that met with in the streets of Liverpool or Manchester. For these nothing is so suitable or economical as granite.

Wood paving requires to be laid in streets having an abundance of sun and air; if placed in confined passages or courts it is kept constantly damp, soon rots, and becomes an insanitary pavement.

The gradient upon which it should be laid should not exceed 1 in 27, as under certain atmospheric conditions it becomes slippery and requires sanding or gravelling. To ensure safety to horse traffic wood pavements must be well laid, with close joints, and kept *scrupulously clean*. Wide joints accumulate dirt, which, with the requisite degree of moisture, becomes dangerously greasy.

As the foundation of the roadway is the real carrier of the traffic, and as the blocks are sure to wear more or less unevenly, they should be made as shallow as is consistent with stability, thus avoiding an unnecessarily large initial outlay.

L

The *advantages* of wood paving include the following points:

(1) It is the least noisy of all known pavements; there is no clatter of horses' feet or rattle of wheels.
(2) It is clean, and manufactures no dust or mud of itself, that which accumulates upon its surface being imported from an outside source.
(3) If kept clean, it affords a good foothold for horses and is much safer than asphalte.
(4) There is little resistance to traction upon it, which, though slightly greater than upon asphalte, is fully counteracted by the better foothold afforded.
(5) It is fairly durable under medium traffic, is moderate in cost, and tolerably easily scavenged and repaired.
(6) It is suited to all gradients up to 1 in 27, and when properly laid has a good appearance.

The principal *disadvantages* which have been urged against wood pavement are as follows:—

(1) That it is an insanitary pavement, owing to its absorbtion of moisture and consequent emanation of offensive smells.
(2) That it is troublesome to open or repair for purposes of access to gas, water or other pipes.
(3) That it is not readily cleansed, except with the aid of water, the dirt adhering to the wood or remaining in the joints.
(4) That the wood expands when it becomes wet, and thus often displaces the curbs.

As regards the alleged insanitary nature of wood pavements, Mr. Lewis H. Isaacs, C.E., surveyor to the Board of Works for the Holborn district, in a paper on the "Construction of Roads and Streets from a Sanitary Point of View,"* observes: "However great may be the claims of wooden pavements upon our gratitude for their noiselessness, it is not only possible, but probable, that we have obtained this undoubted boon at the cost of our producing an atmosphere highly charged with elements not conducive to perfect, or

---

* *Vide* " Journal of Sanitary Institute," vol. xv., p. 144.

even approximately perfect, salubrity. On the score of hygiene a wooden pavement cannot be said to be an ideal pavement, nor can we regard it as the pavement of the future. It will, doubtless, hold its own until superseded by a pavement as noiseless as, but more in compliance with, the sanitary requirements of the present day than itself."

Scientists and medical men have also turned their attention to the question of the failure of wood paving from a hygienic point of view, and have quoted certain organic diseases as being traceable to the large use of wood in cities like London.

These evils, however, have doubtless arisen from pavements which have been improperly laid, with uncreosoted blocks and wide joints. Absorption of moisture is largely prevented by creosoting the blocks, and a similar end may be attained by the use of hard wood and close joints. In regard to the percentage of moisture absorbed, Mr. W. N. Blair, c.e., gives the following figures:*—

Yellow deal blocks	23·44 per cent.
Jarrah blocks	10·71 ,,
Karri blocks	7·77 ,,

A properly laid pavement, constructed of hard wood, with close joints pitch grouted, and kept in good repair and well scavenged and cleansed, may be safely regarded as a sanitary pavement and perfectly free from the insalubrious terrors above mentioned.

Mr. R. W. Richards, c.e., the city surveyor, Sydney, at a meeting of municipal engineers in London, in February, 1897, made the following observations on the question of probable disease germs in wood pavements in Sydney. He said : "I recognised when I first used the wide-jointed wood pavement the danger of its being impregnated with disease germs . . . . and asked to be allowed to commission Mr. McGarvie-Smith, a bacterioligist of great eminence, to examine the grouting taken from the pavement. That gentleman kindly brought sterilised vessels, &c., and collected a

---

* " Proceedings of the Association of Municipal and County Engineers," vol. xx., p. 80.

sample from the grouting of the blocks, and, after having made cultures, he pronounced them perfectly innocuous. Certain growths were specially searched for, particularly typhoid, but he failed to find a single germ. The Board of Health, in 1882, anticipated that pathogenic germs would be found, and that typhoid and other fevers would most likely follow the introduction of wood pavement, but he did not find any germs of diseases pathogenic in man."

WOODS EMPLOYED IN STREET PAVING.

The following woods have been used for paving—beech, oak, larch, elm, pitch pine and Baltic fir; also the following hard woods from West Australia and New South Wales—karri, jarrah, blackbutt, tallow wood, blue gum, red gum and spotted gum. Cedar blocks have also been largely used in America, but are now held in disfavour.

Of the soft woods, the best for the purpose is said to be Wyborg or St. Petersburg red deals, whilst Memel and Dantzic timber is better than Riga on account of the stronger and tougher nature of the former. Swedish yellow deals and American spruce have also been largely used. Pitch pine is very hard, but varies in texture and contains a large amount of resin, which makes the wood very subject to changes of temperature and when laid on damp places soon rots.

In respect to the Australian hard woods, Mr. R. W. Richards (Sydney) gives it as his opinion[*] that jarrah and karri are very good timbers, but that they cannot compare with the blackbutt and tallow wood of New South Wales. Blackbutt, he says, is a timber which has not been properly appreciated, because the Victorian variety has been imported into this country, which is different altogether from that of New South Wales.

"Blue gum is a very good wood, and will hold its own with tallow wood, which is the best. Then come blackbutt and red gum."

"Spotted gum is a treacherous timber, and in Sydney it

---

[*] "Proceedings of the Association of Municipal and County Engineers," vol. xxii

has been found that dry rot soon gets in, and not only consumes this timber but all with which it may be laid. To find good results occurring from the use of this wood is exceptional." Its use has been discontinued in Sydney. "The same may be said of brush box, colonial pine and colonial cedar, all of which have alike been discarded."

The average karri tree is about 200 ft. in height and 4 ft. in diameter at 3 ft. to 4 ft. from the ground, and has its first branches at a height of from 130 ft. to 150 ft. Karri timber is hard, heavy, elastic and tough; the grain of the wood is interlaced, thus giving extraordinary strength.

As to the safety for traffic afforded by karri blocks, Mr. Ednie Brown says: "For street-blocking it is also valuable, and for this purpose seems equal to, if not better than, jarrah, in that its surface does not become so slippery to the horses' feet."

The consulting engineer to the West Australian Government reports that for many uses karri "is superior to jarrah for wood-paving blocks, owing to its not wearing to so polished a surface as jarrah is likely to do."

Karri wood, as supplied to the Fulham Vestry and delivered to their wharf near Putney Bridge, in blocks 9 in. by 3 in. by 4 in., costs 185s. per 1,000, and jarrah 207s. 6d. per 1,000.

The following table (p. 166) is given by Mr. J. P. Norrington, C.E.,* for the comparison of the West Australian woods with English oak and fir.†

The results of experiments made by Mr. Norrington upon karri, jarrah and deal blocks showed karri to sink in water, jarrah to float; also jarrah was shown to absorb more water and to swell more freely than karri.

As to the relative capabilities of karri and jarrah, a commission on the subject, appointed in 1887, arrived at the following conclusions:—

That "it is very certain that the usefulness and durability

---

* "Proceedings of the Association of Municipal and County Engineers," vol. xix., p. 46.

† The figures in the table are obtained from "Timber and Timber Trees," by Mr. Thomas Laslett, timber inspector to the Admiralty.

Description of wood.	Deflection in inches.			Weight in lb. required to break each piece.	Weight in lb. required to break 1 square inch.	Tensile experiment direct cohesion in lb. per square inch.	Crushing strains.		Classed at Lloyds. Years.
	With apparatus weighing 390 lb.	After weight removed.	At crisis of breaking.				In tons.	In tons per square inch.	
Karri ...	1·01	·04	6·06	862	215	7,070	6-in. cube 185	5·14	12
Jarrah ...	3·21	1·33	4·71	685	171	2,910	2-in. cube 12·762	3·198	12
Teak ...	1·65	·83	5·37	912	228	3,301	2-in. cube 11·35	2·838	14
English oak	3·37	·189	7·35	776	193	7,571	2-in. cube 13·625	3·406	9
Fir ...	1·62	·066	5·14	876	219	3,231	2-in. cube 12·687	3·172	—

of both timbers depends very much upon the locality where grown and the season of the year in which it is felled. The commission consider that from November to May or June, or when the sap is down, is the best time to undertake this work; this applies equally to jarrah or karri."

That where severe strains or crushing weight has to be considered, as for the superstructure of bridges, jetties, &c., karri should be preferred and used before jarrah, in consequence of its greater strength.

It did not appear from the evidence before the commission that there is much to choose between the two timbers as far as their ant-resisting properties are concerned. The commission were also of opinion "that karri is most suitable for works of construction, railways, and public works generally; but as regards the timber resisting the action of salt water and sea worms in marine works it has yet to be proved."

The following table (p. 168), also prepared by Mr. J. P. Norrington,* gives the estimated relative cost of using the three woods named.

Karri is therefore cheaper than jarrah, superior in strength and more durable; also karri and jarrah are shown to be both cheaper in the long run than creosoted deal. A hard wood pavement is also greatly superior from a sanitary point of view.

The usual term for which loans are granted for wood pavements is five years; but the county council, recognising the superiority of jarrah wood pavement, offered to lend for seven years to the St. George's Vestry.

In purchasing hard woods for paving the method adopted by the Lambeth Vestry,† which district has over 9 miles of jarrah-paved roads, "is to purchase the timber by the load, the price varying from £5 19s. to £6 2s. 6d. per load, which includes barging alongside the vestry's wharf. It is then unloaded by the vestry's men and sawn into blocks close to

---

* " Proceedings of the Association of Municipal and County Engineers," vol. xix.

† *Vide* "Proceedings of the Association of Municipal and County Engineers," vol. xxii., p. 91.

Description of wood.	Area in square yards	Price per yard.	Total cost.	Probable duration of wood.	Average annual cost of repayment of loan, at 3½ per cent., spread over period of duration of wood.
		s. d.	£	Years.	£ s. d.
Karri wood, 9 in. by 3 in. by 4 in. deep	12,500	10 6	6,562	16	534 12 0
Jarrah, 9 in. by 3 in. by 4 in. deep ...	12,500	11 6	7,187	14	649 5 0
Deal creosoted, 9 in. by 3 in. by 5 in. ...	12,500	7 6	4,687	8	680 0 8

where delivered. It is customary to estimate that 640 blocks, $4\frac{1}{8}$ in. in depth, can be cut from one load, which brings the cost per 1,000 blocks to £9 11s., the cost of cutting being 4s. $9\frac{1}{2}$d. per 1,000. The loss in sawing is precisely $\frac{1}{8}$ in. per block. On testing this seventeen saw-cuts were found to be equal to $2\frac{1}{8}$ in.

" The cutting is done with a single saw, 30 in. in diameter, which is worked by an 8 horse-power gas engine, the saw making 1,200 revolutions per minute; and four saws require to be sharpened each day. The effect is to cut the blocks with great rapidity, thirty-seven blocks being cut in seventy seconds and forty-three blocks in seventy-eight seconds."

On the question of hard wood paving the city surveyor of Sydney, as the result of seventeen years' experience, has expressed the following conclusions*—*viz.*,—

"That a carriageway pavement laid upon a good foundation of concrete with New South Wales hard woods, with slope or butt joints, with convexity of 1 in 60 or 1 in 80, as the longitudinal gradient may suggest, properly cleansed and maintained, is the best and most suitable form of pavement for heavy and continual traffic; and that the best timbers for such work are tallow wood, blackbutt, blue gum, red gum and mahogany. These timbers, after having been so laid in Sydney streets, have, upon examination, shown wear at the rate of from $\frac{1}{50}$ in. to $\frac{1}{30}$ in. per annum, and have not required repairs of any kind whatsoever. The cost of the first wide-jointed pavement, including all labour, materials, &c., was about 27s. per yard super., while recently pavement with the close joints in Sydney cost 15s. 6d. per yard.

"It might be thought that close-jointed pavements set up greater expansion than pavements with wide joints. Experience shows that such is not the case. It may be surprising, but it is nevertheless true, that expansion is, indeed, decreased. To combat any possible expansion, however, a 2-in. seam of sand or clay is laid parallel with and close to the curbing. Outside this layer planks of hard wood 12 ft.

---

* *Vide* " Proceedings of the Association of Municipal and County Engineers," vol. xxiii.

long by 3 in. by 6 in., the full depth of the block, are fixed. These planks are so placed as to break joint with each other on each side of the street, and with such provision expansion has not resulted to any appreciable extent."

CREOSOTING TIMBER.

The best natural means of preserving timber is to have it well seasoned and ventilated. Decay is brought about by the fermenting of the albumenoids of the sap, by the attacks of insects, and by the admission of water into the cells of the wood. Various processes have been employed with the object of preventing decay, by the exclusion of moisture or by expelling or drying up the sap.

These processes consist of "creosoting" and of the impregnation of the timber with metallic salts.

The relative *absorbing power of timber* is given by Molesworth as follows:—

Memel	= 1·00
Elm	= 1·35
Yellow pine	= 1·15
Beech	= 4·00
English oak	= ·34

Creosoting is effected by first extracting the air and moisture from the tubes of the timber and then forcing in *kreasote** at a temperature of about 120 deg. Fahr., and at a pressure varying from 100 lb. to 170 lb. per square inch.

Creosoting about doubles the life of the timber, if properly done, and the cost of treatment does not exceed 20 per cent. of the prime cost of the material. Of all the preservative processes it is the most successful; the carbolic and other acids in the creosote coagulate the albumen of the wood and arrest the process of fermentation; the pores of the wood become filled with the heavier portions of the creosote, which enter, and so exclude moisture. Creosoting also destroys insects and fungi, prevents dry rot and repels worms. Sea worms, and even the white ant, are resisted by creosoting.

---

* *Kreasote*, oil of tar, usually called *creosote;* also spelt *creasote.*

Two of the most common enemies to timber in submarine work is the worm *teredo navalis* and the insect *limnoria terebrans*, the latter resembling in appearance a small wood-louse.

The quantity of creosote forced in will depend upon the nature of the timber and the purpose for which it is to be used. The sapwood absorbs more than the heart. Fir and other soft woods take from 10 lb. to 12 lb. per cubic foot, and beech as much as 24 lb., whilst oak will not absorb more than 6 lb., and red pine 15 lb. Mr. Bethell recommends 7 lb. per cubic foot for railway works and 10 lb. for marine work, but larger quantities than these are now frequently used. Sapwood, fully impregnated, is more durable than heartwood unimpregnated; so that timber with the outer layers intact should be used for creosoting.

Creosote is a product obtained in distilling tar, and is an oily dark liquid, containing hydro-carbons of different degrees of volatility and antiseptic qualities. A heavy oil, well heated and with high pressure, is the most suitable.

Dr. Tidy's specification for creosote required it to be quite liquid at 100 deg., and without deposit until the temperature fell to 95 deg.; also, one-fourth was not to distil over in a retort at less temperature than 600 deg., and this fourth was to be heavier than water. It was to contain 8 per cent. of tar acids by analysis with caustic soda and sulphuric acid, but no shale oil, bone oil, or any oil not distilled from coal tar.

Bone oils and mineral oils are sometimes used for creosoting, but they are unfit for the purpose, as they contain no antiseptics, and should be rejected.

"The minute glistening scales generally observable on newly-creosoted wood consist of napthaline, a substance that possesses considerable antiseptic properties. When this substance exists in the liquor in moderate quantities it thickens and confirms its consistency, but when there is a very large proportion . . . it makes the liquor too solid."*

Crude tar should not be mixed with creosote oils. It con-

---

* "R. E. Journal."

tains all the oils lighter than water, which are useless for creosoting, and a large quantity of pitch, the carbon in which prevents the oils from penetrating the fibres of the timber. Ammoniacal water, if present in the creosote, has an injurious effect upon the timber.

Creosote oils are divided into two classes, known in the trade as London, or heavy oils, and country, or light oils. The London oils are obtained from the gas tar from Newcastle coal, and the country oils from coals from the Midland districts. The London oils contain a large proportion of napthaline, and are heavier and thicker than country oils.

The quantity of tar acids should bear some relation to the quantity of sap in the timber; the tar acids for young or sappy wood should be greater than for deals and whole timbers. About 5 per cent. of tar acids are sufficient for ordinary cases.

The specific gravity of creosote is about 1·05, and 1 gallons weighs 10·5 lb., but this varies with the locality in which it is distilled.

Timber should be inspected before treatment, as it is almost impossible to judge of the quality of the timber afterwards. An inspector should be present during the treatment, to gauge the quantity of oil absorbed, as, except in the case of small scantlings treated by Bethell's process, which can be weighed before and after treatment, there are no means of checking the quantity of oil injected.

Timbers of similar scantlings should be treated in the same charge, so that they all may be equally impregnated. Timber should be prepared before treatment, to avoid breaking through the crust of creosoted fibres by any subsequent carpentering operations.

Creosote is injected into timber either according to Bethell's, Boulton's or Blythe's process. These three processes have been very fully described in a paper[*] on "Creosoting Timber," by Mr. E. J. Silcock, C.E., which descriptions are here adopted:—

---

[*] "Proceedings of the Association of Municipal and County Engineers," vol. xvii.

Bethell's Process.—"The timber is placed in closed iron cylinders, varying in length from 40 ft. to 80 ft., and in diameter from 4 ft. to 7 ft. The cylinders have hemispherical ends, one of which is loose and is suspended from an iron arm fixed to the cylinder, so that during charging and discharging it can be swung on one side.

"The cylinder is provided with a faced flange at the end which opens, and the hemispherical end is secured in its place by screw clamps, which clamp the end of the cylinder flange.

"The creosote is contained in a tank or tanks underneath or to one side of the cylinder, and the tank and cylinder are connected by an iron pipe led from the bottom of the cylinder and provided with a stop valve. Hot-water or steam pipes are taken from a suitable boiler and furnace into the creosote tanks, so that the temperature of the oil can be raised to 120 deg. Fahr. before admission to the cylinder.

"An air pump and a pressure pump, driven either by steam or hydraulic power, are placed in a shed adjoining the cylinder, the air pump being connected to the creosote cylinder by suitable piping, and the pressure pump drawing its supply of creosote from a tank and discharging into the creosote cylinder. It is preferable, if possible, to have a separate tank for the pressure pump to draw from, as the quantity of creosote injected can then be more accurately measured.

"The process is then carried out as follows: The timber is placed in the cylinder, the hemispherical end fixed, and the air pump is set to work to exhaust the air from the space unoccupied by the timber and from the pores of the timber, the valve between the cylinder and the creosote tank being closed and a vacuum of about 10 lb. per inch being maintained. At the end of an hour the valves opened, and the atmospheric pressure causes the creosote to rise up the pipe connecting the cylinder and tank, and fills up all spaces in the cylinder not occupied by timber.

"When the tank is full the air pump is stopped and the pressure pump is set to work, and creosote is pumped in under pressure until the required quantity has been injected into the wood. The time occupied in this operation depends on the quantity to be injected and the state of the timber, varying from ten minutes to as many hours, the pressure varying from 30 lb. per square inch up to 120 lb. per square inch.

"The quantity to be pumped is determined by multiplying the cubical contents in feet of the charge of timber by the specified quantity of creosote per cubic foot. The tank from which the pressure pump draws its supply is provided with a gauge, by means of which the quantity pumped can be measured.

"When the required quantity of creosote has been injected the pump is stopped, and the valve connecting the cylinder with the creosote tank is opened, and the oil is allowed to drain back into the tank. The cylinder is then opened and the charge of timber withdrawn."

Boulton's Process.—This process "differs" from Bethell's in that the temperature to which the oils are raised in the creosote cylinder is 212 deg. Fahr. to 220 deg. Fahr., instead of 120 deg. Fahr. To effect this steam pipes are introduced into the creosote cylinder. The plant required is similar to that for Bethell's process, except that, besides the above-mentioned steam pipes, a steam dome is fitted to the cylinder and a surface condenser to the air pump.

"The principle on which the process rests is the fact that whereas the boiling points of the creosote oils vary from about 250 deg. Fahr. to 700 deg. Fahr., that of water is 212 deg. Fahr., consequently all the moisture present in the timber is driven out in the form of steam by the heat of the oils, and this steam is exhausted by the air pump, and subsequently condensed without any appreciable quantity of creosote being evaporated. At the same time the timber does not suffer from this excessive heat, because that heat is applied through the medium of the oils and is not a dry heat.

"The process is carried out in the following manner: As soon as the cylinder is charged with timber and closed oil is introduced until the cylinder is nearly full, and the temperature raised to 212 deg. Fahr. to 220 deg. Fahr. The air pump is then started, and the air and steam exhausted until no further water flows from the condenser. The pressure pump is then worked until the required quantity of creosote is injected.

"The timber absorbs a large quantity of oils by displacement as the moisture is driven out, so that the pressure pumps have less work to perform, and less oil has to be forced in by that means, a clear advantage over Bethell's process.

"The total quantity of oils injected is the sum of the quantity absorbed by displacement and the quantity injected under pressure. The first of these quantities is ascertained by completely filling the creosote cylinder and dome with oil before the air pump is started, then reading the gauge in the creosote tank.

"A small quantity of oil is then run back from the cylinder into the tank, to secure a space at the top of the cylinder in which the air and steam can gather. When the exhausting process is finished the cylinder and dome are again filled and the gauge again read. The difference between the two readings gives the quantity required—*i.e.*, already absorbed.

"The second of these quantities is the balance required to comply with the specification, and this is then pumped in by the pressure pump from a tank fitted with a contents gauge.

"Boulton's process is a very marked improvement on the older system, more especially when treating timber which has been water-seasoned or which is very green. To put it in plain words, the timber is boiled in the creosote, and by that means all the water and most of the sap are removed, as well as the air from the cells of the timber, and the creosote oils take their place. The advantages of the process are not so marked when treating deals or other sawn timber, which is usually fairly well seasoned and tolerably dry, but it is common practice to take memel timber and pitch pine out of seasoning ponds and put straight into the creosote cylinders.

"In such a case Bethell's process cannot materially reduce the quantity of water in the timber. Merely exhausting the air and reducing the pressure on the outside of the timber will not extract any large quantity of moisture, and when the pressure pump is put to work, and the creosote forced into the timber, the water is only driven more deeply in, and then sealed in by the bituminous portions of the oils. It will thus be seen that it may be quite possible, by using Bethell's process, to fix in the timber an element of danger to its durability."

There is another advantage in Boulton's process. "Having extracted a large portion of the sap in the shape of steam, there is not such a large quantity of albumenoid matter to coagulate; so that the quantity of tar acids may be reduced and oils of an inferior quality may be used without decreasing the efficiency of the treatment. This is important, as the tar acids are now very largely extracted from the creosote oils for other commercial purposes, and with ordinary London oils the percentage of tar acids is not as high as prudence would dictate if all the sap in young timber is left in."

Blythe's Process.—This process is "somewhat similar to Bethell's, except that no air pump is provided and a steam injector replaces the pressure pump." The process, however (known as "Thermo-carbolisation"), differs from Bethell's and Boulton's in principle. "The oils in this system are applied, first in a gaseous or finely divided state, and subsequently as a liquid. The process is carried out as follows:—

"The timber is stacked in a closed cylinder, and superheated steam, at a temperature of 800 deg. Fahr., is supplied to the injector. The injector delivers into the bottom of the cylinder and draws its supply of oils from a tank. There is also a pipe connecting the top of the cylinder with the injector, so arranged that the injector exhausts the gases from the top of the cylinder and redelivers them, along with fresh supplies of finely-divided oil, at the bottom of the cylinder. In this manner a continuous circulation of gases is kept up for thirty minutes, or longer, the temperature of the gases varying from 80 deg. or 90 deg. Fahr. at the commencement to above 212 deg. Fahr. The steam is then turned off, and the cylinder and contents allowed to cool. If the timber is required for joinery or building purposes the process stops at this point; if, on the other hand, the timber is required for underground or hydraulic works, the cylinder is connected with the creosote tanks, and oils are pumped in by steam pressure or ordinary pumps, as in Bethell's process."

It is claimed that this system dries, hardens and protects timber, the oils penetrating the heart as well as the sapwood.

Creosoting in Open Tank.—An imperfect form of the operation of creosoting is sometimes performed by merely steeping the timber in creosote in an open tank. This process requires to be carried on for several weeks, and the oil kept to a high temperature, or it does not penetrate beyond a very limited distance, unless the scantlings are thin or consist largely of sapwood.

The following is suggested as a standard specification for creosoting timber* :—

" The whole of the timber used must be creosoted with coal-tar creosote, which shall be completely fluid at 100 deg. Fahr., shall yield 5 per cent. of tar acids, and contain 25 per cent. of constituents which do not distil under 535 deg. Fahr. The creosoting must be carried out in the following manner : The timber must be placed in a closed iron cylinder fitted with a dome, an air pump, and a pressure pump, and capable of being heated. As much creosote must be admitted to the cylinder as will nearly fill it; the temperature of the oil must then be raised to over 212 deg. Fahr., and the air pump put to work until all moisture, in the shape of steam, has been exhausted. The cylinder must then be completely filled with creosote, and the pressure pump worked until a pressure of 100 lb. per square inch is produced and the timber has taken up creosote at the rate of 1 gallon per cubic foot."

As regards creosoting paving blocks, many surveyors contend that creosoting does not materially increase the life of the pavement when subjected to exceptional traffic, but that the advantage of their use lies in the fact of their being a

* "Proceedings of the Association of Municipal and County Engineers," vol. xvii., p. 221.

M

more sanitary pavement, and in their being better for scavenging purposes—the creosoted blocks drying quicker than the plain ones.

In some districts the objection has been raised against the use of creosoted blocks of the creosote oozing up in the roadways in hot weather, and complaints have been received from the residents of the locality of damage thus caused to their property and furniture.

Creosoting adds about 1s. per square yard to the cost of an ordinary wood pavement.

The process for the preservation of wood known as *Renwickising* consists in boiling the wood in coal tar.

In the processes known as *Seelyising* and *Hayfordising* creosote is also used, the wood in the former case being first boiled, and in the latter unseasoned.

Other processes which have been employed for the preservation of timber consist of the impregnation of the wood with metallic salts; of these may be mentioned that of Boucherie, Burnett, Gardner, Kyan, Margary, Payne.

Boucherie's Process.—In this process the timber is impregnated with sulphate of copper, by filling a reservoir, placed at a height of 25 ft. to 30 ft. above the ground, with the solution (1 lb. of sulphate to 12·5 gallons of water), and conducting the same by means of a pipe into a deep incision into the wood, so that the liquid may reach the centre of the log. The pressure due to the height of the tank enables it to force its way along the sap tubes to the far end of the timber, which upon being rubbed with prussiate of potash will produce a brown stain, which thus affords of proof of the liquid having passed right through the log.

Burnett's Process.—In this process Burnett's chloride of zinc, diluted with from thirty to sixty parts of water, is impregnated under a pressure of from 100 lb. to 120 lb. per square inch for about a quarter of an hour.

Gardner's Process consists in dissolving the sap by chemicals in open tanks, driving out the moisture, and leaving the fibre only. Chemical substances are then injected, which add to the durability and make the timber uninflammable. The process takes from four to fourteen days, according to the nature of the timber.

Kyan's Process.—Corrosive sublimate, diluted with 150 parts of water, is injected. It is said that the process has some effect in retarding dry rot.

Margary's Process.—The wood is soaked in diluted acetate or sulphate of copper.

Payne's Process.—Sulphate of iron, and then sulphate of zinc, are injected into the pores of the wood.

The impregnation of metallic salts for the preservation of wood has not in all cases proved satisfactory; sometimes the fibres of the wood are destroyed.

Wood for paving purposes is usually creosoted by one of the processes above described.

THE CONSTRUCTION OF WOOD PAVEMENTS.

Since the adoption of wood pavements many methods of construction of varying degrees of merit have from time to time been introduced, some of which, however, are now known by name only. The following may be mentioned:—

Asphaltic Wood Pavement.—This system[*] of wood pavement is laid as follows: The ground is excavated and prepared as required; a *foundation* of blue lias lime concrete, consisting of one part lime to six parts ballast, is then laid in to a thickness of 6 in. and to the proper contour of the road section. Upon this is laid a layer of mastic *asphalte* $\frac{1}{2}$ in. to $\frac{3}{4}$ in. in thickness, which then receives the wood *blocks* (3 in. by 9 in. by 5 in., or 3 in. by 8 in. by 5 in. of Baltic fir), which are placed in transverse courses with the grain of the wood upwards. Interspaces, or *joints*, about $\frac{9}{16}$ in. in width are left between the courses, the correct spacing being ensured by placing long strips of wood $\frac{9}{16}$ in. thick against each course as laid. Into these joints melted asphalte is then poured to a depth of about 2 in. up the block, and partially remelts and unites with the coating upon which it rests. The whole pavement thus becomes one rigid mass, and the re-

---

[*] Originally patented by Copland.

maining depth of joint is filled up with a grouting of sand and hydraulic lime, which serves as a non-conductor of heat. A top dressing of fine, sharp sand is then strewn over the pavement, for the purpose of indurating the surface of the wood.

The blocks were originally pierced at the sides with two holes, which were filled with the liquid asphalte to afford an additional key, but in the later pavement of this class they appear to have been dispensed with as an unnecessary refinement.

Carey's Wood Pavement is formed as follows: The paving blocks are cut 4 in. wide by 9 in. long, and to a depth of 5 in. or 6 in., according to the traffic; they are shaped with alternate convex and concave ends, and are laid on a bed of ballast or sand averaging 2 in. The joints are left about ⅜ in. wide, and are filled with a grout of lime and sand.

The advantages claimed for this pavement were, that by the peculiar shaping of the blocks they will better retain their positions and disperse the weight over an extended area. Experience has since shown that the simple rectangular block as now used serves every requirement, and that the labour of converting blocks to special shapes is a useless expense.

Croskey's Wood Pavement.—Mr. Croskey proposed to manufacture cross-grained planks of wood of any length, which, upon being laid side by side, were to be forced together by pressure, and so form a compact homogeneous surface of wood, which was to be laid upon concrete. This plan does not seem to have been tried.

Elli Wood Paving.—This pavement is formed in the following manner: "Excavate the ground 11 in. deep, lay broken stones or gravel 6 in. deep, well ramming same so as to form the proper camber of the street, with a layer of sand ¾ in. deep on top, then place the oak pegs, which are 4 in. long and from 2 in. to 4 in. in diameter, fill up interstices with sand and a small quantity of water, well ram same, sand again, and ram until an even and solid surface is produced."

This pavement has been used in several Continental towns, and has also been tried at Bristol. In respect to the latter place, as may be imagined, an uneven surface resulted from the sand working up between the pegs, which then gave way. An objection to the pavement is its lack of homogeneity of structure.

Gabriel's Wood Pavement.—On a foundation of concrete, consisting of Thames ballast and Portland cement or ground lias lime, were placed wood blocks (3 in. by 6 in. by 7 in. to 11 in. long), resting on a thin cushion of sand. The joints were grouted with lime and sand, and the surface sprinkled with hoggin and sand. In order to keep the blocks steady while grouting two small fillets were secured to one side of each block.

Harrison's Wood Pavement is somewhat similar to the asphaltic wood pavement. Strips of wood 2 in. wide by $\frac{1}{2}$ in. thick are placed upon a concrete foundation, and upon these 3-in. wood blocks are laid, the joints being filled with liquid asphalte, which penetrates under and adheres to the blocks, thus forming an under-coating of asphalte in sections, instead of in large sheets as in the asphaltic wood pavement.

Henson's Wood Pavement.—A coat of ordinary roofing felt, saturated with a hot asphaltic composition, is spread on a foundation of blue lias lime concrete. Upon this, which is intended to give elasticity to the road, blocks (3 in. by 6 in. by 9 in.) of Swedish yellow deal are placed with the grain upright, are driven together, and close jointed with a strip of saturated felt in the joints, which are reduced to less than $\frac{1}{4}$ in. in thickness.

The blocks were bevelled on the top edges, or sometimes at every fourth or fifth block, or a **V** groove was cut down the *centre* of the block, to avoid soakage down the joints.

Improved Wood Pavement.—This was originally formed of two layers of 1-in. redwood boards, previously dipped in boiling tar, laid transversely and longitudinally upon the original road foundation, which was made up with sand on dry earth to the proper contour. Upon these boards the wood blocks (3 in. by 6 in. by 9 in), also dipped in tar, were laid, the longitudinal joints being kept $\frac{3}{4}$ in. apart by a fillet nailed to the flooring, and the heading joints butting. The spaces or joints thus formed were filled with fine dry ballast run with liquid tar and rammed with a flat caulking-iron. The road was then sprinkled with fine gravel.

The object of the flooring of planks was to form an elastic foundation and to tend to distribute the weight; but the use of these has been abandoned since about the year 1877 in favour of a cement-concrete foundation, and other important modifications have also been introduced.

Ligno-Mineral Wood Pavement.—This system appears to have been the first wood pavement laid with a hard concrete foundation. It was introduced from France, and is also known as Trenaunay's system. Upon the concrete bed, which is moulded to the required contour of the road, is laid blocks of hard wood—as oak, beech, elm or ash—which have been subjected to a treatment termed mineralisation and impregnated with hydro-carburetted oils. The blocks (9 in. by 3 in. by 6 in. deep) are cut at an oblique angle of about 60° to the grain, the object being to expose the fibre obliquely to the wearing surface and to distribute the weight of the traffic from the one block to those adjacent in the line of thrust. A horizontal groove is cut on the sides of the blocks near their bases, which is filled with asphaltic mastic (pitch and tar) used for jointing. The upper edges of the blocks are chamfered, and their inclination when laid is alternately to the right and to the left in the successive courses. The joints are only partly filled with the mastic, and are completed to the surface with a grouting of lime and gravel.

In addition to hard woods, mineralised firs are also used, but they are laid upright with vertical fibres.

Ligno-mineral wood pavement was laid in Gracechurch-street, Fore-street and Coleman-street between the years 1872 and 1875.

Lloyd's Patent Keyed Wood Pavement.—Pitch pine blocks grooved on each side are laid direct upon a concrete foundation. Portland cement grouting is run in and forms a key.

Marshall's Wood Paving.—This is a system introduced into the city of Norwich by the borough surveyor, Mr. P. Marshall. Ordinary memel planks are sawn into blocks, and are placed like common paving cubes on a gravel foundation. This kind of pavement has been carried out at prices varying from 5s. 8d. to 7s. 7¾d. per square yard. The pavement is simply laid on the road formation, levelled up with shingle, and the blocks are grouted with blue lias lime and well rammed. The system is said to answer well in Norwich, and makes a cheap wood pavement for country towns.

Messrs. Mowlem & Co.'s Method.—Messrs. Mowlem's method of laying wood pavement is to form a foundation of concrete, varying in thickness according to the nature of the subsoil and the traffic; then to pave with blocks of yellow deal 3 in. wide and 6 in. or 7 in. deep; the joints, which vary from ⅜ in. to ½ in., are filled in with sand and lias lime, and the surface is afterwards indurated by strewing with shingle.

Nicholson's Wood Pavement.—Rectangular blocks of pine are laid upon a close flooring of pine boards 1 in. in thickness. These boards, which are thoroughly tarred, are laid upon a bed of sand lengthwise with the direction of the street, and rest at their ends upon similar boards laid at right angles to the line of street from kerb to kerb. The joints of the blocks are run with asphaltic mixture, and the surface covered with hot coal tar and strewed with sand and gravel.

Norton's Wood Pavement.—Upon the subsoil is laid a thin bed of ballast, which is brought to the profile of the street to receive the pavement. This consists of slabs (7 ft. by 3 ft.), made up of yellow Baltic timber blocks (3 in. by 6 in. deep by 7 in), attached to a backing of 2-in. boarding with a strong bituminous cement, which also fills the joints between the blocks. The joints, ¾ in. in width, between the slabs are filled with powdered rock asphalte in a heated state. The object of this pavement seems to have been to dispense with a concrete foundation.

Prosser's Wood Pavement consists of blocks sawn at an angle of 60°., the grain of the wood being in the same direction. A plank of the same depth as the blocks is placed between the different rows, and the blocks on one side lean in the opposite direction to those on the other, and are dowelled together by wooden pins passing through the plank and entering the blocks about 1 in.

Shiel's Composite Block Paving.—Composite blocks (12 in. by 15 in.) are cast in iron moulds, with two rows of wood fixed at equal distances from each side and each other. The spaces thus made are filled with broken granite, and over this is poured boiling composition of pitch, chalk and sand. The blocks, when cool, are taken to the street, laid upon a concrete foundation, and cement grouted.

Stone's Wood Pavement.—Grooves $\frac{3}{4}$ in. deep and 3 in. apart were cut by machinery in the surface of a concrete foundation, and wood blocks (3 in. wide, 5 in. deep, and 6 in. long), shaped to fit the grooves, are placed on the concrete. The pavement was laid with $1\frac{1}{4}$-in. joints filled with gravel and run in with heated tar.

Stowe's Wood Pavement.—In this pavement, which is of American origin, the blocks are laid upon a 6-in. bed of sand or gravel. " The blocks are set in courses transversely across the street, so as to break joint lengthwise of the street, the courses being separated from each other 1 in. by a continuous course of wooden wedges placed close together edge to edge, and extending from curb to curb. These wedges are set in the first instance with their tops flush with the top surface of the blocks. After the whole pavement shall have been well rammed, so as to give each block a firm bed, the wedges are driven down about 3 in., and the open joints thus formed above them between the courses are filled in with a concrete composed of hot coal tar and fine roofing sand and gravel. The surface of the pavement may then be coated with coal tar prepared by boiling with pitch, and finished off with a thin layer of sand."*

Wilson's Wood Pavement consists of blocks (3 in. wide by 6 in. deep by 8 in. long) laid upon a blue lias lime concrete foundation 6 in. thick. The blocks are bedded upon a $\frac{1}{2}$-in. layer of sand, and the joints are grouted with blue lias lime

---

* "Practical Treatise on Roads, Streets and Pavements," by Q. A. Gilmore.

and sand. To keep the blocks steady before grouting two small fillets are braded to the side of each block.

In constructing a wood pavement experience has long since shown that, as in most other things, the simplest mode of construction is the best. Patent interlocking, grooved or dowelled blocks, and special elastic foundations, have all proved themselves unnecessarily complicated, and have therefore been abandoned.

A good wood pavement, as laid at the present day, consists essentially of the following : A rigid foundation of Portland cement concrete 6 in. thick, and floated over to an even surface, conforming with the contour line of the proposed finished roadway. Upon this, when sufficiently set, rectangular wood blocks, 9 in. by 3 in. by 6 in., cut die square with the fibre vertical, are laid with close joints and pitch-grouted. To allow for the expansion of the wood transversely across the street an expansion joint is provided next the curb. The interstices between the blocks are sometimes grouted with liquid Portland cement and fine sand, necessitating a somewhat wider joint. The whole surface of the pavement is finally spread with a coating, about ½ in. in thickness, of fine sharp gravel or chippings.

If soft woods are used the blocks should be creosoted, as, although the life of the block is not materially increased thereby, the pavement is rendered much less absorbent and more sanitary. The preparation known as "*Carbolineum Avenarius,*" which is largely used by railway companies for preserving sleepers, has also been used as a preservative for paving blocks. The blocks are immersed, without pressure, in the preparation for from ten to fifteen minutes at a temperature of from 125 deg. to 135 deg. Fahr. The cost of treatment is about the same as creosote.

Wood blocks should be laid with the fibre of the wood quite vertical. The following results of experiments upon this point are given by Mr. H. Percy Boulnois[*] :—

---

[*] "Carriageways and Footways," by H. Percy Boulnois, M.I.C.E. (Biggs & Co.)

Vertical fibre		wore	·125	of an inch.
Laid at angle of	75°	,,	·147	,, ,,
,, ,,	60°	,,	·182	,, ,,
,, ,,	45°	,,	·250	,, ,,
,, ,,	30°	,,	·310	,, ,,
,, ,,	15°	,,	·375	,, ,,
Horizontal			·500	,, ,,

A *cushion bed* of sand, ½ in. in thickness, as shown in Fig. 39, is frequently laid upon the concrete foundation. This bed absorbs the bitumen grouting which reaches it, and so tends to prevent the upheaval of the blocks which occasionally occurs where the blocks are laid directly upon the surface of the concrete.

Joints in Wood Pavements.—A good deal of difference of opinion has existed on the question of joints in wood pavements. The customary practice until recently has been to have a ¼-in. or ⅜-in. joint between the blocks grouted with Portland cement and sand. The form of joint used by the Improved Wood Pavement Company is to first run in hot tar and pitch to a depth of 1 in. or 1½ in., and then to grout the rest of the joint with cement and sand.

It is the practice of some engineers to pave wood blocks as close as possible, even using sledge-hammers to drive the blocks together, whilst others adopt a ⅜-in. joint. There is also a divergence of practice in the grouting material; it may be cement and sand, lias lime and sand, a bituminous mixture of tar and pitch, or tar alone, or pitch and creosote oil. Sometimes coal tar is used and sometimes Stockholm tar.

In modern wood pavements joints between the blocks are dispensed with and the blocks are laid close together, touching at the sides and ends. Hot tar is poured over the surface and "squeegeed" or brushed until it disappears between the blocks. The surface is finished by sprinkling fine grit, shingle or gravel, over the pavement and allowing the traffic to squeeze it into the surface of the wood, thus preserving its life and rendering the wood less slippery. An expansion joint (see Fig. 39) along the curbs provides for the expan-

187

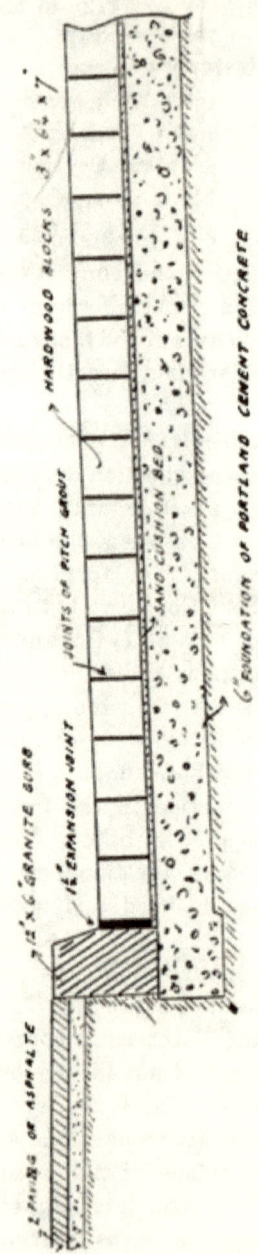

Fig. 39.

sion of the blocks, which, though comparatively small with hard woods, is very considerable with the softer woods, amounting in some instances to as much as 1¼ in. in 8 ft. One inch to every 10 ft. has been allowed for an expansion joint in the channels, but no fixed rule can be made as to expansion, as it varies with almost every pavement. An expansion joint of 1 in. or 1½ in. on each side of a street 60 ft. wide has been found sufficient in Bristol, where karri wood has been laid. Great trouble was caused by this expansion in the earlier wood pavements, which often displaced curbs, lamp-posts, &c., and damaged gully gratings and other fixtures.

It is not necessary to specify wood for paving purposes to be "perfectly well-seasoned." It has always to occupy a somewhat damp position, and is better if it contains a certain amount of "life" in it. It is preferable that the hard woods should be laid as soon as possible after introduction into this country.

From figures* ascertained by Mr. W. N. Blair (St. Pancras) it appears that "Australian woods shrink and expand to a considerable extent under the conditions of dryness and saturation, but at the same time their capacity for absorbing water is much less than that of yellow deal; indeed, in this respect they compare favourably with a good gault brick, which is rarely found to absorb less than 10 per cent. its weight in water. It is due to this variation in dimension that such trouble is often caused in wood-paved roads, necessitating either the cutting out of a channel course or the resetting of curbing and footway paving; and all this more than once, for the action is repeated so long as the pores of the wood are not permanently filled." This is more particularly the case with the softer woods, but the Australian hard woods are also subject to this expansion. In the case of "a roadway 36 ft. wide, paved with karri blocks, with ⅜-in. joints, grouted with cement and sand, laid between 12-in. granite channel blocks, the curbs and footways have

---

* "Proceedings of the Association of Municipal and County Engineers," vol. xx.

twice been laid, and now need it again, and all within two years of the first construction of the road."*

In another case "a carriageway, 39 ft. wide, was paved with jarrah blocks, with ¼-in. and ⅜-in. joints grouted with cement and sand. When the road was being paved a 3-in. joint next the curb was formed by first laying a 3-in. deal along, which was removed as soon as the grouting was set. In four days this joint was taken up; then a channel course was taken out and split, but expansion continued until one 4 in. on each side of the road was measured, and that seems to have been the maximum."*

On the other hand, in roads paved and jointed in the following manner Mr. Blair has found no case of pressure being exerted upon the curbs and the channel joints have not been reduced in width: "Roads, varying from 22 ft. to 48 ft. in width, laid with yellow deal, pitch pine, jarrah and karri, with ⅛-in. and $\frac{1}{16}$-in. joints, regulated by a lath of that thickness and 1 in. wide, laid on edge in the bottom of the joint and left there, the cross-joints being left open the same width but without the lath. The channel joints have been from ¾ in. to 1½ in., and the whole of the joints filled with a hot compound of pitch and creosote oil, proportioned suitably to obtain a freely ductile mixture at a temperature of about 60 deg. Fahr., the proportions being usually 1 cwt. of pitch to 1½ or 2 gallons of oil, some variation being necessary to meet the differences of quality of the pitch. This compound is poured over the paving out of a bucket, and a rubber squeegee is used to push forward the fluid, so that as little as possible remains on the surface."*

Instead of laths, 3-in. strips of tarred felt have been used in the joints to keep an ⅛-in. space between the courses with satisfactory results; but the cost of the felt was about 3d. per square yard, whilst the laths cost less than 1d. per yard super.

In the parish of St. Marylebone yellow deal blocks are laid as close as they can be driven, and flushed over with hot tar, a method which has produced some excellent pavements.

---

* "Proceedings of the Association of Municipal and County Engineers," vol. xx.

Two men with sledge-hammers drive the blocks together at every twelve or fifteen courses. The cross-joints are not perfectly close, but are left, to provide for expansion.

In Chelsea the method has been tried of dipping each block in a box containing a hot pitch mixture, and immediately placing it close up to the adjoining blocks, and then flushing over the surface with hot pitch.

Where wide joints are used the fibre of the wood is allowed to spread and disintegrate, which greatly reduces the life of the wood. This evil increases in proportion to the width of the joint. Wide joints also are the cause of noisy traffic, and they retain a large amount of dirt, which increases the expense of scavenging and gives discomfort to the public using the streets.

Although cement or lime grout for wood pavements is very largely employed in London, both Manchester and Liverpool use pitch without exception, and this form of jointing is increasing in general favour.

Cement Grout requires a joint about $\frac{1}{4}$ in. wide, thus allowing the arrises of the courses to become worn, and so forming a "corduroy" road, which has the attendant disadvantages of becoming noisy and dirty, whilst the life of the blocks is also reduced. Cement, too, is a rigid material, and is not capable of absorbing the expansion of the wood blocks, which movement is consequently transmitted to the channels, resulting probably in the displacement of the curbs, &c., as previously referred to. A carriageway laid with cement-grouted joints requires to be closed a week, so that the cement may properly set and the joints not disintegrated by the traffic. The joints may be made in any weather, which, of course, is an advantage towards facilitating the progress of the work. A rut is frequently to be observed next the curb, caused by the traffic wheels running on the edge of the channel course.

Pitch Grout allows the joints to be close, so as to prevent the spreading of the fibre of the wood. The blocks may be in contact, or not exceeding $\frac{1}{16}$ in. in width, thus preserving better surface. Pitch joints also are of an elastic nature, which admits of their absorption of the expansion of the wood, thus obviating the transmission of lateral pressure upon the curbs. The small expansion joint next the curb does not vary much in width after the completion of the work. There is the disadvantage in the use of pitch joints that they can only be made on dry material, and when the weather is wet the blocks must be covered with tarpaulins; but as soon as the joints are run the traffic may be allowed to pass over. The blocks of cement-jointed pavements sometimes become loose in the jointing material, and the pavement is therefore no longer impervious, but pitch-jointed pavements are always impervious to moisture and the pitch preserves the blocks. The cost of cement grouting is about 6d. per super. yard, and of pitch grouting 7d. per square yard.

The pitch and creosote oil for jointing are boiled together in a travelling pitch boiler. Healey's patent pitch and tar boilers, manufactured by the Municipal Appliances Company of Preston, are suitable for this work. When boiling the pitch is drawn off into a bucket containing about 10 gallons, and is carried on a rod by two men to the pavement and run into the joints. A long-handled ladle or pitch scoop is sometimes used to fill the bucket. In Liverpool a better plan is adopted—viz., of drawing off the pitch from the boiler into a wheeled pitch carrier holding 20 gallons. This portable carrier is provided with a double lining filled with a nonconducting material which retains the heat.

In Liverpool, where there is an agitation for wood pavement from the abutting owners instead of granite setts, "wood pavement is laid if the owners prepay 60 per cent. of the estimated first cost of the wood pavement, the council maintaining it for ten years, reserving to themselves the right to substitute granite setts for the wood at the end of that period if they should so decide."[*]

A good form of contour for wood-paved roads is that adopted by Mr. Deacon, of Liverpool, and which is given in a paper read before the Institution of Civil Engineers in 1878.

---

[*] "The Construction of Carriageways and Footways," by H. Percy Boulnois, M.I.C.E.

This contour (Fig. 40), the sides of which are flatly curved and the crown more so, approximates to a hyperbola. The ordinate

$$R^0 = \frac{a}{36} \text{ or } \frac{a}{45}$$

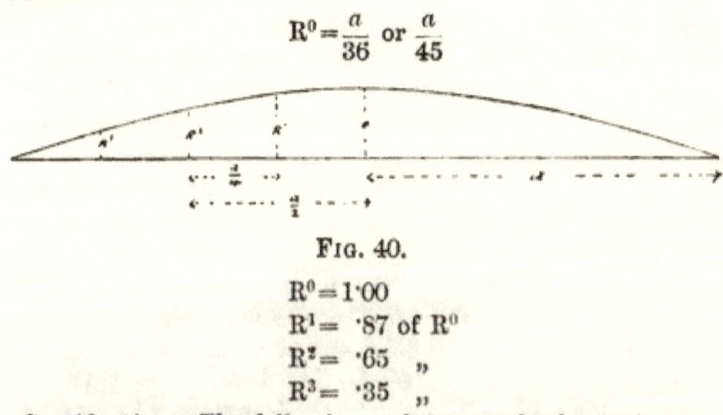

Fig. 40.

$R^0 = 1\cdot00$
$R^1 = \cdot 87$ of R^0
$R^2 = \cdot 65$ „
$R^3 = \cdot 35$ „

Specification.—The following points are of primary importance in specifying for the construction of wood paving:—

Excavate the ground to the required level below the proposed finished surface of the carriageway; the excavated road material, if macadam and approved by the surveyor, may be screened and used as concrete in the foundation. Any surplus macadam, as well as granite pitching, in crossings or channels thus excavated should remain the property of the authority. The formation service should be watered, rolled and punned, and any soft earth removed. Sufficient notice should also be given to gas, water, electric or other companies to enable them to attend to their mains and services before the foundations are proceeded with.

The foundations should consist of not less than 6 in. of good Portland cement concrete, composed of one of cement to two of fine sharp sand and three of broken stone. The concrete must be finished off to an even surface—of a contour similar to that of the proposed finished carriageway. About ½ in. of fine sand is then spead over the concrete surface, and upon this the blocks are laid.

The wood blocks must be of the best Swedish or other approved yellow deals (or of the best Baltic red timber, of karri or jarrah wood, or of such other wood as may be required), free from shakes or sap, loose or dead knots, of a close and uniform grain, and to be equal to three sample blocks deposited at the surveyor's office. It should also be required that any blocks that may be rejected by the surveyor must be at once broken up or removed from the works. The blocks should not be more than 12 in. or less than 6 in. in length, and should be 3-in. in width and 6 in. in depth. They should be carefully laid with close joints, with the fibre of the wood placed vertically, and grouted with either pitch or cement grout (as desired) in the manner above described.

The surface of the wood pavement is then to be sprinkled with a $\frac{1}{2}$-in. coating of fine sharp gravel or chippings.

The contractor must be required to complete the work within a given time and to maintain it for a specified time after completion; also to at once make good any damage caused to water, gas or other mains or services.

Any alteration necessitated to street gullies, sewer manholes, lampholes, &c., such as raising or lowering to suit the new finished surface, are usually specified to be done by the authority's own workmen.

Payments are made to the contractor as the work proceeds up to, say, 80 per cent. of the work executed, and the remaining 20 per cent. paid at the end of about two years after completion.

In making good wood pavements after they have been disturbed by water, gas or other companies for the purpose of repairing or laying mains, additional paving should be taken up and the concrete foundation exposed for a distance of not less than 1 ft. all round the excavation. Loose material should be dug out, and the excavation filled in and thoroughly well rammed. The edges of the concrete foundation should then be sloped back, and good Portland cement concrete filled in to the required thickness, say 10 in. for a 12-in. foundation, the remaining 2 in. to consist of floating of Portland cement and sand in the proportion of one to three by measure. After the expiration of four clear days the blocks may be laid and grouted with cement and sand, and the traffic allowed to pass over at the end of three days after repaving. Every care is necessary in this work to avoid subsequent settlement.

Wear of Wood Pavements.—This necessarily depends upon a number of circumstances peculiar to the position in which a pavement is laid, such, for example, as the amount and weight of the traffic, the gradient of the carriageway, &c. To form a true comparison of the effect of traffic upon pavements its approximate weight and the width of the street must be taken into account, and these particulars then reduced to a standard, such, for example, as so many tons weight of traffic per yard of width of carriageway per annum.

The life of an ordinary wood pavement may be taken as from eight to ten years. The amount of annual wear was estimated by Mr. Deacon at from $1\frac{3}{8}$ in. to $2\frac{5}{16}$ in.* In a report upon asphalte and wood pavements, by Mr. William Haywood (1874), it is stated that "wood pavements with repairs have in this city (London) had a life varying from six to nineteen years, and that with repairs an average life of about ten years may be obtained."

The life of the hard woods is considerably longer than that of the timber used before their introduction, but, as no hard wood pavements have yet been worn out in this country, it is impossible at present to state what their life will be.

In the early part of 1893 a portion of the western end of Euston-road was repaved in sections, with the object of testing the durability of yellow deal, jarrah and karri. The vestry surveyor, Mr. W. Nisbet Blair, C.E., states:† "The *wear* upon these different lengths has been recently measured and may be taken as at three years from the date of laying. It proves to be $\frac{1}{4}$ in. on the jarrah, $\frac{1}{4}$ in. on the karri, and $1\frac{3}{8}$ in. on the deal blocks. These measurements were taken in the centre line of traffic, where the wear might be regarded as greatest, and at about the centre of each length of wood. . . These figures are equivalent to $\frac{1}{12}$ in. per annum on the jarrah and karri and $\frac{1}{2}$ in. on the deal, and the relative rates of

---

* "Proceedings of the Institution of Civil Engineers," vol. lviii., p. 82.
† "Proceedings of the Association of Municipal and County Engineers," vol. xxii.

wear are as one to six." A seven-day record of the traffic was kept, and was found to be equal to 575,544 tons per yard in width per annum. During the week's record 7·72 per cent. of the total was omnibus traffic. Another length of the Euston-road, at St. Pancras Church, paved with jarrah blocks in 1894, has been under observation, and the wear ascertained to be ·18 in. per annum under a traffic of 411,318 tons per yard in width per annum. In this case 31·8 per cent. of the total weight was omnibus traffic. The wearing effect of omnibus traffic upon the surface of a roadway is believed to be much greater than that of general traffic; this must particularly apply to omnibus stopping places.

Karri wood laid on the Bristol bridge, which crosses the river Avon in the direct line to the railway stations, showed about $\frac{1}{3}$ in. wear at the expiration of twelve months.

In Sydney (New South Wales) different classes of wood pavements, laid on a 6-in. concrete foundation with 1-in. joints filled in with bluestone screenings and tar, wore as follows*:—

Blue gum	$\frac{1}{10}$ in. per annum.
Mahogany	$\frac{1}{8}$,, ,, ,,
Turpentine	$\frac{1}{11}$,, ,, ,,
Brush box	$\frac{1}{7}$,, ,, ,,
Spotted gum	$\frac{1}{7}$,, ,, ,,
Baltic	$\frac{1}{10}$,, ,, ,,
Colonial cedar	$\frac{1}{12}$,, ,, ,,
Black butt	$\frac{1}{20}$,, ,, ,,
Colonial pine	$\frac{1}{22}$,, ,, ,,
Red gum	$\frac{1}{10}$,, ,, ,,

Later carriageway pavements in Sydney of New South Wales hard woods, laid with slape or butt joints on a concrete foundation, with a convexity of 1 in 60 or 1 in 80 (according to the longitudinal gradient), have shown a wear at the rate of from $\frac{1}{80}$ in. to $\frac{1}{80}$ in. per annum, without repairs of any kind. The best woods are tallow wood, black butt,

---

* "Proceedings of the Association of Municipal and County Engineers," vol. xxiii.

blue gum, red gum and mahogany. The cost of the first-mentioned wide-jointed pavement, including all labour and material, was 27s. per square yard, whilst the more recent close-jointed pavement cost only 15s. 6d. per square yard.*

The experience with these close-jointed pavements is that, contrary to what may be anticipated, expansion is decreased. To obviate any possible trouble in this direction, however, a 2-in. seam of sand on clay is laid parallel with and close to the curbing. Outside this are fixed planks of hard wood, 12 ft. long by 3 in. by 6 in., the full depth of the blocks. These planks are laid so as to break joint with each other on each side of the street, and it has been found, with this provision, no appreciable expansion has resulted. The traffic in some of the principal streets in Sydney is from 350 to 500 vehicles per hour. Each vehicle, as an average, may be taken as representing 1 ton in weight.*

ASPHALTE PAVEMENTS.

Rock asphalte is a natural rock, consisting of pure carbonate of lime impregnated with mineral bitumen in variable proportions, that for carriageways and footways containing from 7 to 12 per cent. of bitumen.

The rock is quarried in a similar manner to other materials, and breaks with an irregular fracture without definite cleavage. The grain should be regular, homogeneous and fine. That principally used for carriageway pavements in Europe is derived from Val de Travers in Switzerland, Pyrimont Seyssel in the Jura mountains, Sicily, Limmer and Vorwohle in Germany. The Limmer Asphalte Paving Company's supplies for *compressed* work are taken from Ragusa in Sicily, and for *mastic* work from Germany. The mineral is brought to England in its natural state.

The rock found at Seyssel is a limestone saturated with bitumen, containing about 90 per cent. carbonate of lime and 8 to 10 per cent. of bitumen.

* "Proceedings of the Association of Municipal and County Engineers," vol. xxiii.

The Val de Travers asphalte is from a rock found at Neuchatel, and is said to be richer in bitumen than Seyssel asphalte, containing from 11 to 12 per cent.

Too little bitumen in asphalte for paving purposes renders it brittle and liable to crack, and, on the other hand, too much makes it liable to the action of the heat of the sun. That containing more than 10 per cent. of bitumen becomes soft in summer, and if containing less will not bind properly. Asphalte from different mines should not be mixed.

What is known as *compressed* asphalte is used for carriageways, and that termed *mastic* is only suitable for footways, courtyards, &c., of light traffic.

The manager of the Limmer Asphalte Paving Company has given the following particulars of the process of manufacture of *compressed* asphalte for carriageways :—

The pieces of raw rock, weighing from ¼ cwt. to ½ cwt., are placed in the "crusher," where they get broken to the size of walnuts.

After passing the "crusher" the material is carried by elevators to a disintegrator, which runs at a speed of about 1,800 revolutions per minute, where it is ground to a fine powder.*

For compressed work this powder is heated in roasters with revolving cylinders to a temperature of about 280 deg. Fahr., and as soon as the superfluous moisture has evaporated the heated powder is placed in iron-sheathed vans, covered with cloths, and taken to the street where it is to be laid. The powder should not lose more than 20 deg. of heat during transit.

The powder is then raked over the surface of the street to the thickness required, and rammed with hot pelons, smoothed to a polished surface with a curved hot iron tool, and then rolled with a heavy hand roller weighing about 1,100 lb. It gradually becomes cold, and in time is compressed into solid rock again by the traffic.

The following are the usual thicknesses of compressed asphalte :—

* To pass a sieve of 0·1 square inch mesh.

Carriageway of heavy traffic, 2 in.; carriageway of light traffic, 1½ in.; laid upon a bed of Portland cement concrete 6 in. thick.

Footways may have 1 in. of compressed asphalte laid upon a bed of Portland cement concrete 3 in. or 4 in. in thickness.

The *mastic* asphalte as used for footways, &c., is manufactured in the same way as the compressed as far as the powder stage; then it is boiled in cauldrons, worked by agitators, with a certain quantity of bitumen as flux, and when heated to about 400 deg. Fabr. a proportion of from 10 to 20 per cent. of fine Bridport grit is added and thoroughly incorporated with the asphalte by the agitator; it is then turned into iron moulds and made into cakes of ½ cwt. each, ready to be sent out.

These cakes are broken up, put into cauldrons with just sufficient bitumen as a flux, and when properly cooked the asphalte is spread over the surface, to be covered by means of hand floats, and then rubbed with fine sand.

The bitumen used in the manufacture of mastic asphalte is a most important feature, and its manufacture requires great care. The cheaper kinds are made from petroleum refuse, gas tar, &c., but these materials have a deleterious effect on asphalte. The best bitumen is made from best Scotch shale and refined Trinidad pitch (*épurée*). It is very expensive, but it is of the greatest importance to have it of the best and purest possible quality, in order to render the asphalte capable of standing extremes of heat and cold.

In laying a compressed asphalte roadway care must be taken that the surface of the concrete is perfectly dry before the asphalte powder is laid, as any moisture thereon will be sucked up into the powder, converted into steam, force its way through the powder, and so form fissures, which may only appear sometime after the laying, thus leading to the disintegration of the mass and consequent breaking up of the material.

The iron punners used in the ramming process should weigh about 10 lb., and be heated to prevent the adhesion of the asphalte powder. The ramming is done lightly at first, to ensure equality in thickness of the material, and subsequently increased to heavy blows.

A light sprinkling of sand is spread over the surface when the material is compressed, and a few hours after the consolidation the carriageway is ready for the traffic.

In streets having tram lines it is not advisable to lay asphalte close to the rails, but to here substitute granite setts.

The specific gravity of natural asphalte varies in direct proportion to the quantity of bitumen; the average is 2·25, and a cubic yard weighs about 3,874 lb.

Asphalte carriageways have many advantages, among which may be mentioned the following:—

(a) It is impervious to moisture, and is pre-eminent as a *sanitary* pavement. It is jointless, has a smooth, even surface, and is therefore quickly and easily cleansed; it may be *washed*, and so kept scrupulously clean to a degree such as no other pavement will admit.

(b) It is durable, quickly laid and readily repaired, thus affording a minimum inconvenience to traffic. When properly executed it shows less signs of repairs after gas, water or other pipes than any other class of pavement.

(c) It is comparatively* noiseless, and admits of great ease of traction.

(d) Pedestrians may use both the carriageways and the footways.

(e) It is very pleasing to drive over, there being a minimum of vibration; the pavement is also of an agreeable and uniform colour, and cellars, vaults, &c., under the streets so paved are, owing to its impervious nature, kept dry, and thus rendered more serviceable.

(f) It neither absorbs nor radiates heat, nor gives off emanations, as is sometimes alleged of wood pavements.

The one objection to asphalte is that in certain states of the atmosphere it becomes slippery. Horses slip considerably at the commencement of a shower of rain, when the dirt which has been allowed to remain on the surface of the carriageway becomes a little damp and makes a greasy

---

* The clatter of horses' feet on asphalte is often very apparent.

surface until the pavement becomes thoroughly wet and cleansed.

Owing to this want of foothold it is also found difficult to stop a horse drawing a heavy load, and, when stopped, it is sometimes painful to see the struggles of horses endeavouring to set their loads in motion again. A fallen horse has also great difficulty in rising again; but, on the other hand, the animal does not break his knees in falling upon asphalte. Horses are nervous in passing from one class of pavement to another, owing to the varying degree of foothold afforded.

Sand is sometimes sprinkled upon asphalte carriageways to render them less slippery, but there is the objection to the sand cutting into the surface and defacing the asphalte. The best plan is to keep the roadway thoroughly cleansed by frequent washing; the objection to this is the expense of water and labour.

Asphalte should not be laid upon gradients greater than 1 in 60.

Cost.—The cost of compressed asphalte roadways depends very largely upon local circumstances; in London the first cost per square yard is about 12s. 6d. The following comparative statement of the cost of various road surfaces, compiled by Mr. Ellice Clark, c.e., in 1879,* will be of interest. The cost is there given at 18s. per square yard, but since that time the first cost has been considerably reduced.

Description.	Original cost per square yard.	Interest on Original cost.	Sinking fund at 3 per cent, compound interest.	Maintenance per square yard.	Scavenging per square yard.	Total.
	s. d.	d.	d.	s. d.	d.	s. d.
Val de Travers compressed asphalte	18 0	9·7	—	0 3·6	0·4	1 1·7
Wood (Liverpool)	15 1·5	7·5	10·1	0 1	2·7 gravel 5·0	2 2·3
Granite	17 9	9·6	·5	0 1·3	2·5	1 1·9
Macadam	4 9	2·1	—	3 6	12·0	4 8·1

* "Proceedings of the Association of Municipal and County Engineers," vol. vi.

In the latter the cost is reduced to 100,000 tons per annum per yard of width. Nothing is charged for renewal of asphalte, as the annual sum for maintenance provides the asphalte in perpetuity.

It has been the practice of the asphalte pavement companies to charge a fixed price per square yard for laying, according to the thickness required, the distance from London, &c., and also to undertake to keep in repair at an agreed price per square yard per annum for a certain number of years, as frequently is also done in the case of wood pavements. In lieu of this, repairs may be arranged for under a schedule of prices according as the circumstances may require.

A customary period for maintenance by the contractors is seventeen years, and the agreed cost of maintenance per square yard for the contract term is usually two years free, and fifteen years at from 6d. to 10d. per annum, according to circumstances.

The durability of asphalte is greatly enhanced by its elasticity, and compressed asphalte begins to *wear* only when compression has been completed, which under ordinarily heavy traffic takes two years, and a square foot of the pavement at the end of that time will weigh nearly the same as when first laid, although the thickness may be reduced in the two years as much as it will be in the next eight.

Too great care cannot be taken with the concrete foundation, to have it strong enough to resist the traffic. Thorough dryness is also absolutely essential, as the best asphalte laid upon damp concrete will soon be disintegrated from underneath by the moisture being sucked up and converted into steam by the hot asphalte. The steam in its escape through the heated powder forms fissures, which appear as soon as the surface begins to wear.

The asphalte must be regarded simply as a veneer or covering for the concrete to withstand the *wear* of the traffic, whilst the concrete foundation is solely responsible for carrying the *weight* of the traffic.

NOTES OF A SPECIFICATION FOR COMPRESSED ASPHALTE ROADWAYS.

The ground must be excavated to the required depth, and any defective or soft places well rammed, filled in, and made solid, and a foundation of not less than 6 in. of *good* Portland cement concrete laid in and floated to a smooth and true contour.

The natural rock asphalte is to be crushed in a stone-crusher, and afterwards pulverised in a disintegrator to a fine powder, capable of passing through a sieve of ·1 square inch mesh. The powder is then to be heated to between 240 deg. and 250 deg. Fahr. in revolving cylinders, and afterwards transported in iron covered carts to the street where it is to be laid, and must not lose more than 20 deg. of heat in the course of transit.

If the concrete is perfectly dry the asphalte powder may then be spread in an even layer, $2\frac{1}{2}$ in. in thickness, and carefully raked to ensure regularity of depth and surface.

It is then to be well rammed with iron punners of 10 lb. weight, and heated to avoid their adhesion to the powder. The ramming is to be done lightly at first and afterwards augmented to heavy blows. The pavement is then to be smoothed with suitable curved hot iron tools, and again well rammed and rolled until quite cool. After a few hours the carriageway will be ready for traffic, having been first sprinkled with a little sand.

Test.—The following test[*] for asphalte will be of interest:
" A specimen of the rock, freed from all extraneous matter, having been pulverised as finely as possible, should be dissolved in sulphurate of carbon, turpentine, ether or benzine, placed in a glass vessel, and stirred with a glass rod. A dark solution will result, from which will be precipitated the pulverised limestone.

"The solution of bitumen hould then be poured off. The dissolvent speedily evaporates, leaving the constituent parts

---

[*] "Minutes of Proceedings of the Institution of Civil Engineers," vol. lx.

of the asphalte, each of which should be weighed so as to determine the exact proportion.

"The bitumen should be heated in a lead bath, and tested with a porcelain or Baumé thermometer to 428 deg. Fahr. There will be little loss by evaporation if the bitumen is good, but if bituminous oil is present the loss will be considerable. Grilled mastic should be heated to 450 deg. Fahr.

"The limestone should next be examined. If the powder is white and soft to the touch it is a good component part of asphalte, but if rough and dirty on being tested with reagents it will be found to contain iron pyrites, silicates, clay, &c. Some asphaltes also are of a spongy or hygrometrical nature. Thus as an analysis which merely gives so much bitumen and so much limestone may mislead; it is necessary to know the quality of the limestone and of the bitumen.

"For a good compressed roadway an asphalte composed of pure limestone and 9 to 10 per cent. of bitumen, non-evaporative at 428 deg. Fahr., is the most suitable.

"Asphaltes containing much more than 10 per cent. of bitumen get soft in summer and wavy; those containing much less have not sufficient bind for heavy traffic, although asphalte containing 7 per cent. of bitumen, properly heated, does well for courtyards, as it sets hard when cold."

FOOTWAYS.

Width.—The width of footways, like so many other matters in road construction, must be regulated largely by local requirements, and must be such as may be suitable to meet the exigencies of the trade and traffic of the neighbourhood. Speaking generally, in this country the width of the carriageways has been excessive as compared with that of the footways. The footpath on each side of the road may advantageously be one-fifth of the total width of the roadway, but there may, of course, be exceptional cases where the vehicular traffic is sufficiently numerous to require an additional width of carriageway. The width of the footway,

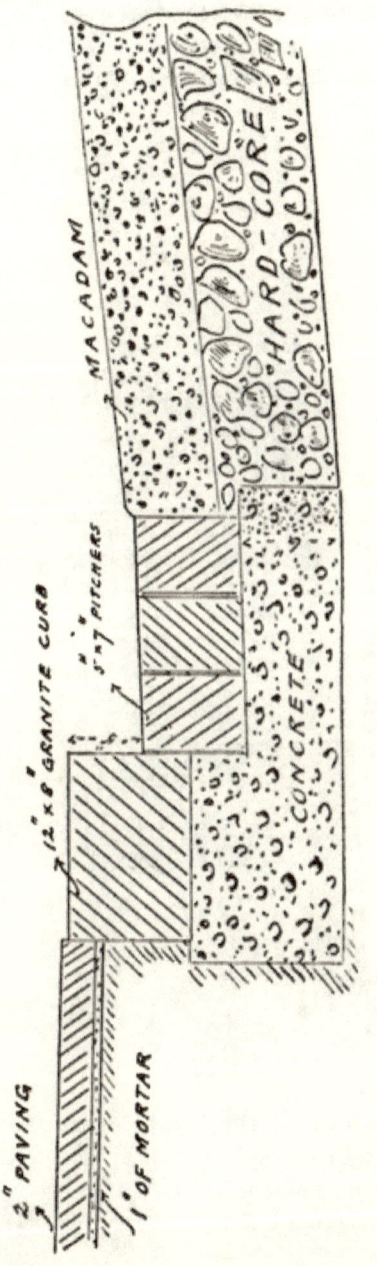

Fig. 1.

as compared with that of the carriageway, should be in about the same ratio as the accommodation required by the pedestrian and vehicular traffic respectively.

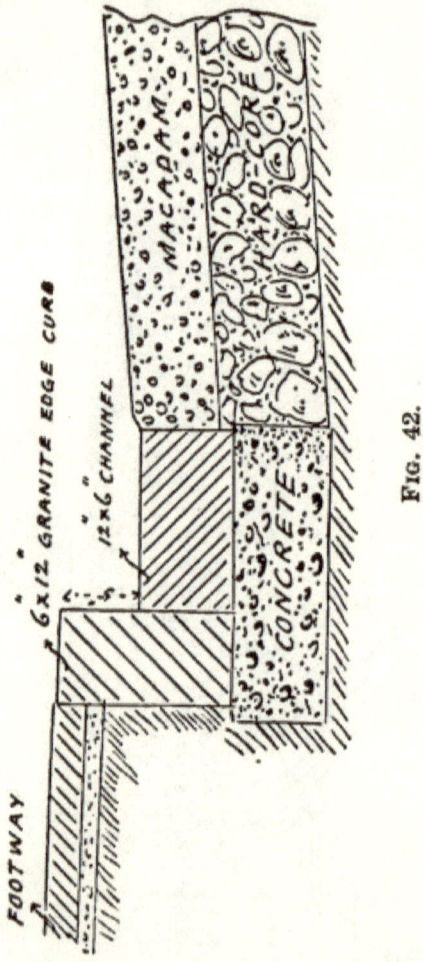

Fig. 42.

The Model By-laws of the Local Government Board prescribe that the person laying out a "new street" "shall construct on each side of such street a footway of a width not less than one-sixth of the entire width of such street;"

also that the footway shall "slope or fall towards the curb or outer edge at the rate of ½ in. in every foot of width if the

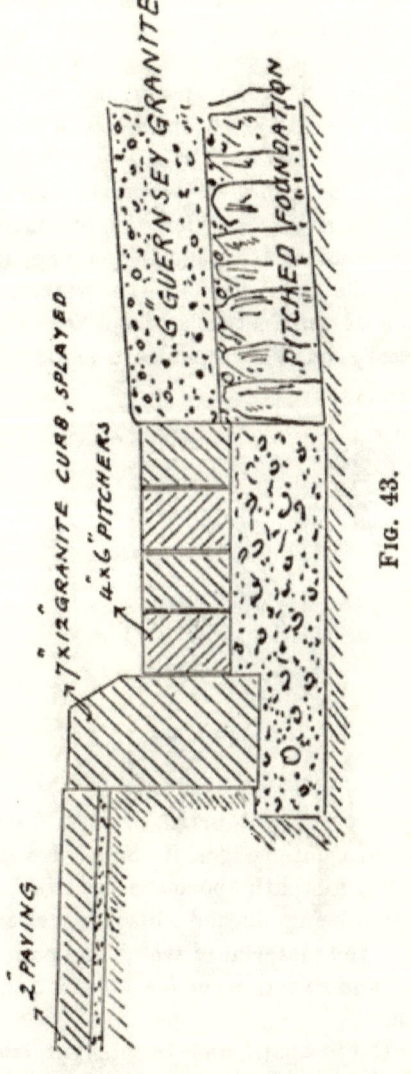

Fig. 43.

footway be not paved, flagged or asphalted, and at the rate of not less than ¼ in., and not more than ½ in., in every foot of width if the footway be paved, flagged or asphalted."

The footway, too, is to be so constructed that "the height of the curb or outer edge of such footway above the channel of the carriageway (except in the case of crossings paved or otherwise formed for the use of foot passengers) shall be not less than 3 in. at the highest part of such channel and not more than 7 in. at the lowest part of such channel."

The object of this latter requirement is that a height of less than 3 in. would be likely to allow vehicles to drive on to the footways, or to admit of water in the channels overflowing the footpath. If the height exceeded 7 in. it would be inconvenient for pedestrians and also render the curb liable to tilt over toward the gutter. For natural stones a slope of the footpath of not less than ¼ in. to the foot is necessary, but for asphalte a slope of about ⅛ in. is sufficient.

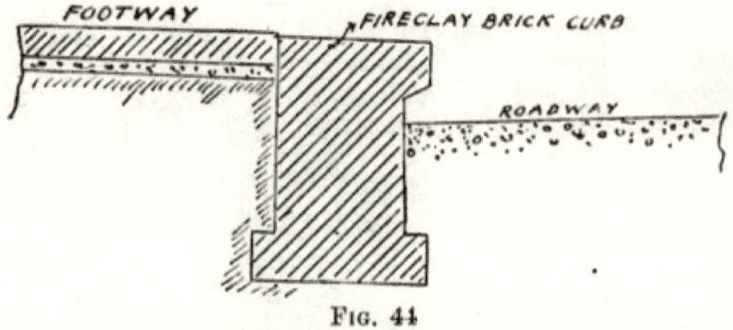

Fig. 44.

CURBING AND CHANNELLING.

All footways, whether in urban or rural districts, require a curb fixed at their outer edges, its object being to—

(a) Raise the footpath above the roadway, and so prevent it from being flooded; also, to act as a sill against which the material of which the path is formed may abut, and so retain the foundation and surface of the path.

(b) Define the footpath, and to prevent traffic driving on to it.

(c) Form the side of the gutter, and to enable the haunches of the roadway to be properly finished.

Material for Curb.—The materials employed for curbing are Aberdeen, Guernsey, Norway or other granites, syenite, sandstone, vitrified fire-clay blocks, cement concrete, cast and wrought iron.

Granite Curbing.—Granite or syenite are by far the best materials, and should always be used in streets having much

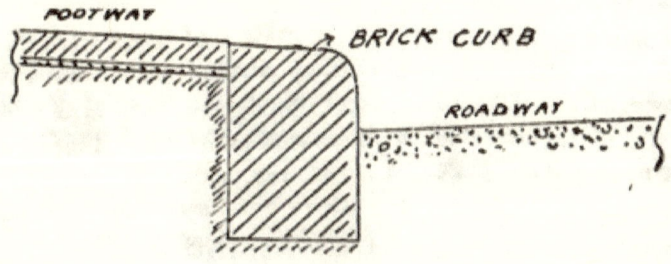

FIG. 45.

traffic, as the curb is always more or less subjected to rough usage from the passing of heavy vehicular traffic, and the grinding action of the wheels of heavily-laden carts, drags, waggons, &c. On gradients it is a very general practice

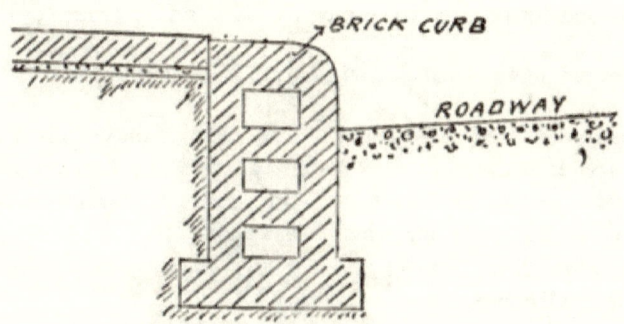

FIG. 46.

among drivers and carmen to "hug" the curb with their vehicles, so as to act as a drag or brake, and the material of which the curb is composed must therefore be capable of withstanding this treatment.

The stone which stands best is Aberdeen granite, but as Norway granite is delivered into the London market at a lower price, very little Aberdeen stone is now used in and about London. Cornish and Newry granites are also used, but are coarse-grained, and neither are well dressed. Guernsey granite is very hard, becomes slippery, and can only be procured in short lengths.

Curbing is also made of Endon or Yorkshire stone, of Purbeck, Keniton and Pennant limestones, but are not to be compared with granite.

Cost of Curbing.—The cost of curbing and channelling necessarily varies largely, according to the locality, nearness to quarries, the value of labour, &c., but the following are the approximate prices :—

Per lineal yard.

Norway granite curb, 12 in. by 8 in. (laid flat), in London ... ... 6s. 6d.
Granite channelling, consisting of three courses of 4-in. by 6-in. pitchers on 6 in. Portland cement concrete, laid in London ... ... 6s. 0d.
Granite curb, 12 in. by 8 in. (bevelled), in Plymouth, 3s. 6d.
Limestone curb, 12 in. deep by 4 in. wide, from { 2s. 6d. to 3s. 6d.
Limestone channel, 10 in. wide by 6 in. thick, about 3s. 6d.

Size and Fixing of Curbing.—The usual dimensions for a curb are 12 in. by 8 in. laid flat, 6 in. by 12 in. laid on edge; it should never be less than 9 in. in depth, narrower than 4 in. in width, or in lengths of less than 3 ft. When 8 in. or more in width at the top the surface should be bevelled to conform with the slope of the path.

The curb should be hammer-dressed, about 5 in. on the face, over the whole surface of the top, and for 3 in. at the back; it should also be drafted about 1 in. along both top edges. This gives a clean appearance to the aris of curb inside and out, presents a smooth surface against the channel gutter, and enables the paving to abut fair against it at the back. Figs. 41, 42 and 43 show various forms of curb and channelling. Where traffic is heavy its grinding action against the curb is minimised by splaying off the outer face, as shown in Fig. 43.

A flat curb is more liable to be displaced than an edge curb, and unless the curb is omitted for lamp-posts the lamps project a good distance on to the footpath. A flat curb, however, affords a substitute for paving where the gravel footway is allowed to get out of repair.

Fireclay Brick Curbs (Figs. 44 to 47) are successfully employed in America, but are not much used in England.

Concrete Curbs.—These are useful for streets of very light traffic, but must not be used where they would be likely to receive heavy blows or grinding action of passing traffic. They are made either *in situ*, or, better, in blocks made in moulds. The concrete should be of the best materials,

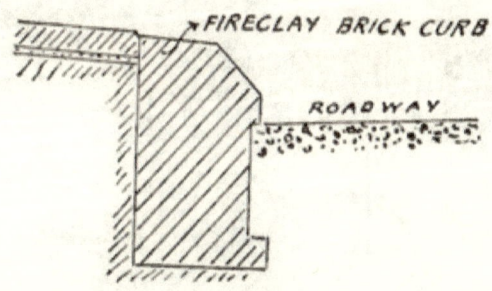

Fig. 47.

including good Portland cement and thoroughly clean shingle, ballast, or broken stone, in proper proportion and well mixed.

Wrought and Cast-Iron Curbs.—Wrought and cast-iron curbing has been used in France and other places, but must wear very slippery, and become a source of danger to pedestrians. Two sections are illustrated in Figs. 48 and 49.

Setting Curb.—Curbing requires to be set by an experienced mason possessing a good eye, in order to make it appear graceful as regards line and level. It must be set carefully and well bedded; otherwise it will sink or tilt over. The "skillet line" and "boring rods" are used to assist in securing a good and pleasing line, both as regards level and

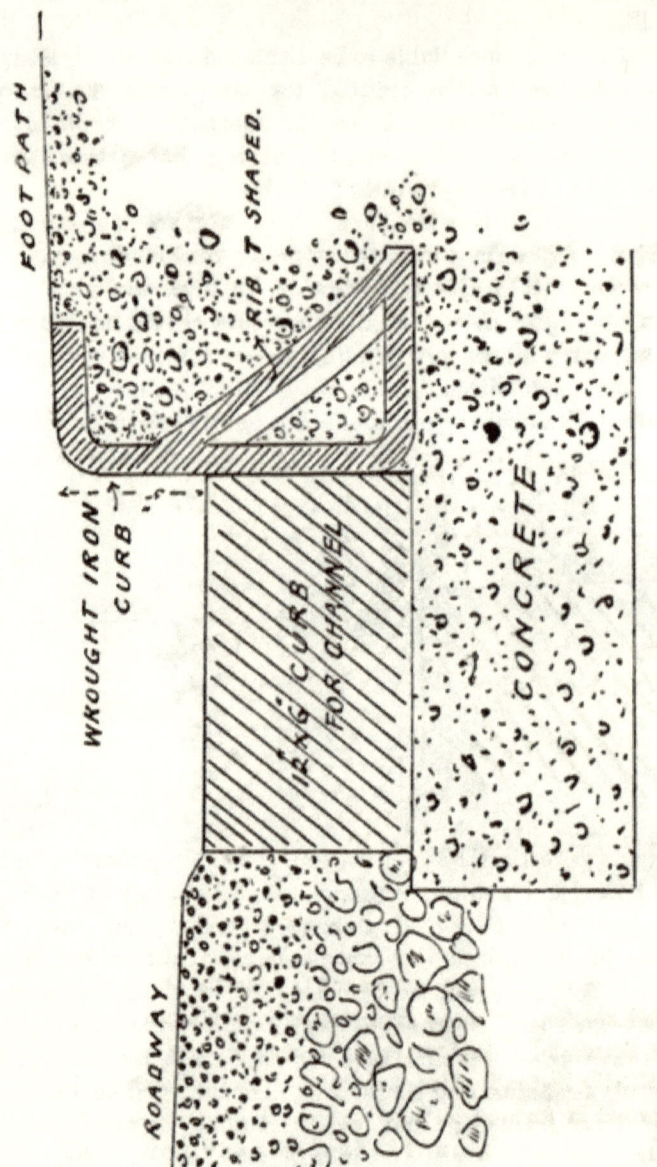

Fig. 48.

contour; but unless the mason possesses a good eye some rather unsightly effects are often produced.

Channelling.—Channels (see Figs. 41, 42 and 43) are necessary to receive the water from the roadway and to convey it away to the gullies, and thus effect satisfactory drainage to the road. A channel also prevents the curb being undermined by water running at the roadsides. Channelling should not be less than 12 in. wide and not more than 18 in.; it may consist of from three to six courses of granite pitchers, or of flat granite slabs about 15 in. wide by 4 in. thick, both laid upon concrete. The approximate prices of channelling have already been given above.

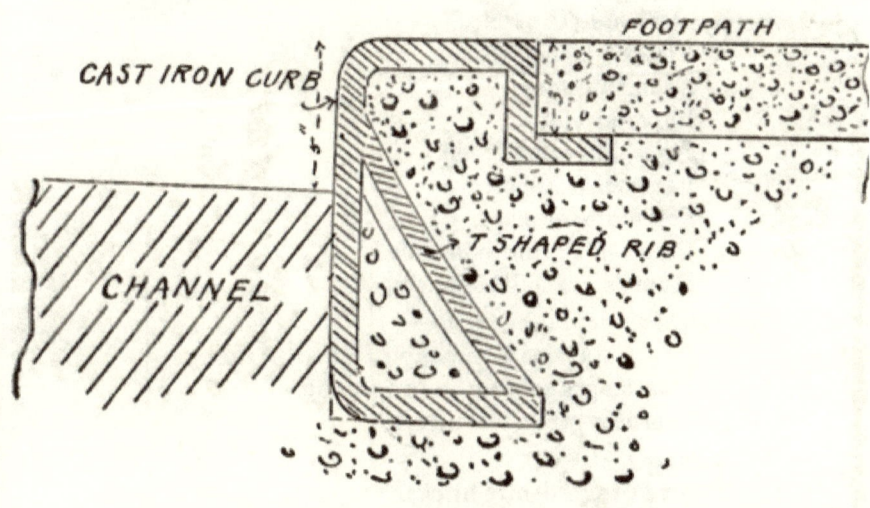

FIG. 49.

Where streets are paved with granite setts, asphalte, or wood blocks, the same material is employed for the channels, the setts or blocks forming the channelling being laid longitudinally instead of transversely, as in the street itself. The channel is formed to take the slope or cross-contour of the adjoining roadway, and is fixed so as to show about 5 in. of curb above it. Channelling must, of course, be laid to fall to the gullies, and must not be flatter than about 1 in 300.

MATERIALS USED FOR PAVING FOOTPATHS.

The classes of material used in paving footpaths are very varied and differ largely, according to the locality and facilities for obtaining same. The following are some of the principal materials used :—

Natural Stones—
    Granite slabs.
    Yorkshire and Caithness flagging.
    Pennant flagging.
    Purbeck flagging.
    Blue lias and Devonian limestone flagging.
    Lazonby flagging.

Artificial Stones (Concrete Slabs)—
    Victoria stone.
    Adamant stone.
    Imperial stone.
    Jones' annealed stone.
    Ransome's stone.
    Ferrumite stone.
    Bucknell's granite breccia.
    Stuart's granolithic paving.

Concrete Paving, laid in situ—
    Imperial Stone Company's *in situ* paving.
    Stuart's granolithic *in situ* paving.
    Guernsey granite concrete.

Brick Paving—
    Blue Staffordshire bricks.
    Imperishable granite vitrified bricks (Candy & Co.).
    Parry's Buckley bricks.

Tar Paving—
    Various varieties and modes of laying.

Asphalte—
    Compressed, as also used for carriageways.
    Mastic, as used for footways only.

Qualities for a good Paving Material.—A good paving material requires to be durable, non-abrasive, smooth, but not slippery. It should absorb the minimum amount of water, be easily repaired, and strong to resist heavy blows from falling articles. It should be uniform in texture, so as to wear evenly, and should possess the combined qualities of efficiency and economy. The surface should be such that dirt will not readily adhere to it, and the colour should be uniform.

Before selecting a stone it is best to lay a short section of footpath with the material where the amount of the traffic is known, and then to note its wear, &c. Its absorption may be gauged by soaking the stone in water and noting the increase in weight.

NATURAL STONES USED FOR PAVING.

Granite Slabs.—Granite slabs, from 3 in. to 6 in. in thickness, are sometimes used in positions where the traffic is exceptionally heavy or where heavy weights are likely to be let fall upon the footway. Granite forms an exceptionally durable pavement, but for general use would in most cases be very expensive, heavy to handle, difficult to work, and in course of time wears slippery.

Yorkshire Flagging.—This is very largely used, and possesses many of the essentials of a good paving material. It affords a first-class foothold, is pleasant to walk upon, and can be easily taken up, redressed or turned, and re-used. It does not wear slippery, and the stones are of large size and can be made to properly break joint. Its appearance is agreeable, and it can be easily worked to irregular frontages, cellar openings, &c.

On the other hand, it has important disadvantages. The surface of a footpath paved with Yorkshire flagging wears unevenly; hard and soft stones will be found side by side, and pools of water will be found upon the pavement. There is also a tendency to laminate in frosty weather. The stone requires to be well bedded to prevent water accumulating underneath, which, when the stone is trod upon, oftentimes squirts out in an objectionable manner to pedestrians. The first cost of Yorkshire flagging in some districts is also high as compared with its durability.

Yorkshire flagging (laid) costs :—

Per square yard.
Ashton-under-Lyne (tooled Rochdale, 3 in. thick), 5s. 4d.
Birmingham, 2½ in. thick ... ... 5s. 7d.
Chelsea, 3 in. thick ... ... 8s. 7d.

In Bradford, Yorkshire flags can be laid at 3s. 6d. per square yard.

The following are prices (1891) of different thicknesses of York paving delivered at any railway station in London:—

	3 in., per sq. yard.	2½ in., per sq. yard.	2 in., per sq. yard.
White, or grey and blue, best and hardest quality, special tooled or polished	6s. 9d.	6s. 6d.	5s. 3d.
White, or grey and blue, ordinary tooled	5s. 9d.	5s. 6d.	4s. 3d.
Brown quality, ordinary tooled ...	5s. 6d.	5s. 3d.	4s. 0d.

The price *laid* would be about 1s. per square yard extra.

York stone is divided into two classes—*viz.*, "soft-bedded" and "riving stone." The former is very rarely used for flagging in London, and the latter is again subdivided into white, or grey and blue, and brown.

The white (or grey and blue) is the hardest and most durable quality, and is obtained from the lowest beds of the Halifax district.

The brown comes from a higher stratum, is a softer class of stone, and easier to quarry and work. It is also more subject to lamination.

A very large proportion of the footways in Kensington are paved with Yorkshire flagging, 3 in. or 2½ in. thick, edged with 12-in. by 8-in. dressed granite curb. Mr. W. Weaver, C.E., the surveyor, in regard to the respective merits of the various materials used for footway pavements is of opinion that "there is nothing like good hard York, *provided it is good.*"[*] Mr. Weaver also considers it an economical pavement, and as bearing upon this fact he says, "It may be mentioned

---

[*] "Proceedings of the Association of Municipal and County Engineers," vol. xviii.

that 3-in. York laid twenty-five years ago, at a cost of 10d. per foot, on the south side of Kensington-broadway, and since that time never repaired, was taken up last autumn (1891); it was then re-faced, re-squared, and re-laid in a subsidiary street, at a cost of 3d. per square foot. It will last in its present position at least another twenty-five years, and will then be taken up, and the best of it worked up and laid in narrow courses round pleasure gardens, &c.; the residue will be sold partly to builders, at 2d. or 3d. per foot (according to size), for coring, and the remainder used for road foundation. It will thus be seen that York has several lives, and when, in a large district like Kensington, several miles of footway have annually to be renewed, at a cost of about £2,500 per mile, it behoves the local authority and its advisers to be careful in selecting the material which gives the best and most economical results, considering the whole cycle of its user. Another point not to be disregarded is the adaptability of the material to the incessant and increasing disturbance for gas, water, electric lighting, telegraph, and other purposes."

SPECIFICATION FOR YORK PAVING.

In specifying for Yorkshire flagging the following points should be kept in mind (The suggestions may, of course, be also readily adapted to any class of pavement, according to circumstances.) :—

The stone to be not less than 3 in. in thickness, and of the very best quality, from Halifax, the quarries in the neighbourhood of Bradford, or in Yorkshire, and subject to the approval of the engineer. To be of a hard and even texture, free from shakes, flakes and laminations, and to be chisel-dressed to a fair and true face, out of winding, properly squared, and not pitched off only or undercut, but to hold good to the square. The flagging is to be properly and neatly cut to receive all water or electric boxes, gratings, coal shoots, &c.

The joints are to be set flush, bedded and pointed with the best blue lias mortar, and there must not be more than fourteen pieces to the 100 square feet.

The bed for the flagging to be made with suitable earth, dry rubbish, or gravel, and all surplus rubbish or earth is to be carted from the site as it arises in the carrying out of the works.

The contractor must watch and light the works, provide all necessary guards, and be responsible for any damages to gas, water, or other pipes, or to any other property, whether public or private, and make all good to the satisfaction of the engineer or parties concerned. He is also to provide all stone, materials, implements, tools, tackle, &c., also all labour, and to complete the work within the time specified to the entire satisfaction of the engineer. The work to be measured up upon completion, and payment to be made as the work proceeds to the value of 80 per cent. of the work executed. The contractor to keep the whole in perfect repair for six months after completion.

Caithness Flagging.—This is the only natural stone which can compete with Yorkshire flagging as a paving material. It comes from Thurso (Scotland), and possesses many excellent qualities.

It has great durability, wears evenly over the whole surface, and does not become slippery, or scale or flake, is impervious to wet, dries rapidly after rain, and is not affected by frost. It can be re-used when half-worn, is economical, and can be laid from $1\frac{1}{2}$ in. to 2 in. in thickness. Natural faces can be used, thus effecting a saving of labour; the edges are sawn, and the joints can thus be expeditiously and well made. It is a cleanly pavement, and dirt and dust do not adhere to it. Caithness stone gives greater resistance to bending stress than Yorkshire flagging, and vehicular traffic may cross it without injury to the pavement.

Pennant Flagging.—This is a sandstone obtained at Fishponds, near Bristol, and is largely used for paving footways. It is of a dark blue and grey colour, and weighs 168·2 lb. per foot cube.

Purbeck Flagging.—Purbeck is a limestone found in the Oolitic series in Dorsetshire, and is much used for paving. It is of a brownish-grey colour, and weighs 169·2 lb. per foot cube.

It is very durable, but wears rather slippery. It is usually in small stones.

Blue Lias and Devonian Limestone Flagging.—Blue lias makes a cheap, durable and clean footway, and has other good qualities, but sometimes wears slippery. Devonian limestone is largely used in the West of England.

Lazonby Flagging.— Lazonby stone and flags constitute part of the Penrith Sandstone, which forms the lowest member of the Permian group in the district around Penrith (Cumberland). The Penrith Sandstone is an immense mass of orange-red and yellowish sandstone, and, besides being the principal source from which local building stone is obtained, yields excellent flags. These flags have been used in Carlisle from time immemorial, are very durable, and give good foothold when wet or dry. Some of these flags have been in use upon the Carlisle footways over sixty years, and in that time have been taken up, re-dressed and re-laid. The city surveyor says,* " Good, selected, smooth, self-faced Lazonby flags of first quality, when laid on a good foundation on a 1-in. bed of ground mortar, will wear evenly down from 3 in. to $1\frac{3}{4}$ in. in thickness or less."

ARTIFICIAL PAVING STONE.

Patent Victoria Stone.—This is composed of finely-crushed granite from Groby, in Leicestershire (washed by patent machinery), and Portland cement, carefully selected, manipulated and moulded, and subsequently steeped in a patent solution of natural soluble silica, by which it is hardened and rendered practically non-porous. The following is the process of manufacture† :—

" Three parts of aggregate are thoroughly mixed in a dry state by machinery, and the water then added in a careful manner, so as to avoid the danger of washing out any of the fine and more soluble portions of the cement, and before

---

* "Proceedings of the Association of Municipal and County Engineers," vol. xiv.

† See "A Practical Treatise on Natural and Artificial Concrete," by H. Reid (E. & F. N. Spon).

any initial set of crude concrete mixture can arise it is put into the moulds, in which it is worked with the trowel so as to fill up the angles and sides, thus ensuring accurate arrises all round . . . . The moulds are made of wood, being lined internally with metal, not only to secure accuracy of form, but also to render them durable and proof against the liability to distortion incidental to the varying character of the work they have to perform. Slabs are made of various sizes, but it is found that the most useful sizes for London paving work are 2 ft. 6 in. in length by 2 ft. wide and 2 ft. square by 2 in. thick. Paving slabs of these sizes weigh 25 lb. to 26 lb. the foot super., and are convenient to handle.

"The moulds, filled in the manner thus described, are allowed to remain on the benches of the moulding-sheds until the concrete has sufficiently set and so much of the water of plasticity evaporated as will permit the slabs to receive the beneficial influence of the silicating operation. This indurating process is one of absorption, and the best practice is that which provides a reasonably porous mass, to which may be introduced an accurately-prepared liquid silicate of the desired specific gravity. The slabs, when sufficiently dry, are relieved from the surroundings of the moulds, which, being made in pieces, can be readily detached by unscrewing the fastenings. The slabs are then taken to the tanks in the silicating yard (protected from the weather), placed side by side, and covered by the silicate solution, where they remain until the proper beneficial influence has been duly imparted. The period of time required to complete the silicating process is not of a fixed or arbitrary character, and depends on the condition of a slab and its capacity of absorption. About fourteen days, under ordinary circumstances, is regarded as sufficient to secure the desired advantage of the process.

"The slabs, after being taken from the tanks, are stacked in the stone-yard, where they remain to season, and are taken away in the order of their age.

"The analysis of a piece of Victoria stone paving is: Silica, 50·35; alumina, 11·87; oxide of iron, 7·33; lime,

18·33; magnesia, 2·03; potash, 1·78; soda, 3·81; carbonic acid, 1·80; water, organic matter, &c., 2·70—total, 100·00.

The crushing weight per cubic inch of Victoria stone is 8,321 lb.,* and the tensile strain per square inch is 1,310 lb.

The stone when laid has a pleasing appearance and makes a neat-looking pavement. It wears uniformly, does not become slippery, and has great durability.

Victoria stone paving in the vicinity of London costs about 6s. per square yard laid. A second quality of stone is also manufactured by the Victoria Stone Company, which they term *indurated stone*; this costs about 5s. per square yard laid.

Adamant Paving Stone.—The process of the manufacture of artificial stone by the Adamant Stone and Paving Company differs from that of other makers in that the method adopted is a mechanical operation and not a chemical process. Instead of relying upon silicate of soda, the materials, when properly mixed in a suitable mould, are subjected to an enormous hydraulic pressure—a slab of 18 in., for example, receiving a pressure of 500 tons. This effectually disposes of both air cavities and moisture, and renders the amalgamation of the materials complete, and produces a dense, non-porous stone.

The stone is composed of finely-crushed Aberdeen granite and the finest Portland cement. It can be manufactured, ready for use, in a few days, and in cases of emergency slabs have been forwarded and laid within one week.

The ordinary sizes of slabs for paving are—

3 ft. × 2 ft. ⎫
2 ft. 6 in. × 2 ft. ⎬ 2 in. in thickness.
2 ft. × 2 ft. ⎭

From a test † of the crushing weight it appears that patent Adamant stone, 1 square foot, 2½ in. thick, was crushed at 413·1 tons.

The stone has been largely used in Wimbledon, Leyton, Wandsworth, Putney, Fulham, Lambeth, Plumstead, West

---

* D. Kirkaldy & Son.
† D. Kirkcaldy & Son   August 12, 1891.

Ham, Tottenham, Hornsey, Brighton, Aberdeen, Edinburgh, and other places.

In the neighbourhood of London this class of paving costs about 5s. 3d. per square yard, laid.

Imperial Stone Pavements.—Imperial stone is made from granite broken to pass a $\frac{3}{16}$-in. sieve, screened and washed, mixed with three parts, by measure, with one part of best cement. Soluble silica is used for the induration of the stone, and is a clear, viscous substance, made from pure flint and caustic soda which are digested by heat under pressure in Papin's digester on a similar apparatus. The materials are thoroughly incorporated, in a dry state, in a horizontal cylinder by machinery, and water afterwards sparingly added and the mixing continued. The concrete is put into metal-lined moulds with well-defined arrises, and then placed on a "trembler"—a machine giving a rapid, vertical, jolting motion to the mould. This motion gets rid of many of the air spaces, and is said to produce a more uniform slab than can be made by hand. The whole operation of mixing the concrete and making the slab in the mould is completed in six minutes. The slabs are taken from the moulds at the end of a two-day rest and are air-dried from seven to nine days. They are then immersed in a silica bath for seven or nine days more, and are afterwards stacked in the open for some months before use.

Among other artificial stone pavements which have been introduced, and more or less extensively used, may be mentioned: "Granolithic" pavement, Ransome's artificial stone, Bucknell's granite "Breccia," "Ferrumite" and others.

Paving slabs are also very frequently manufactured by town surveyors from the clinker arising from the burning of town refuse in refuse furnaces.* These slabs are not, of course, by any means equal to those made from more durable material, such as granite, but in second or third rate streets they form a useful and cheap pavement as regards first cost.

---

* For full details on this subject see "The Removal and Disposal of Town Refuse," by William H. Maxwell, C.E., assistant engineer and surveyor, Leyton (published by the Sanitary Publishing Company, 5 Fetter-lane, E.C. Price 15s. nett).

Where shingle or other suitable material is readily obtainable surveyors frequently make their own concrete pavements, which can be done and laid for from 2s. 6d. to 3s. per square yard. Fine Portland cement concrete is well rammed into wooden moulds lined with iron to ensure a true arrise, and oiled to prevent adhesion, and when sufficiently set the moulds are taken to pieces and the slabs placed in a bath or stacked in the open air. If well bedded these slabs will withstand considerable shocks, and when worn they may be taken up and relaid with the other face upwards in streets of less traffic

CONCRETE PAVING LAID "IN SITU."

The employment of concrete for footpaths has increased during recent years, and improvements have been effected in the mode of laying. The material was originally laid with large, exposed surfaces, which were effected by changes of temperature, resulting in the pavement being damaged by large cracks and hollow arches upon its surface.

This class of pavement is now laid in bays about 6 ft. wide. Each bay is completed alternately, and the intermediate one allowed to set before the adjoining one is commenced. Sometimes laths or strips of soft wood are placed between the widths of concrete, and these widths sub-divided by cutting into the concrete before it is set with a trowel, by which means the concrete is split up and thus allowed room to expand.

In laying concrete footways care must be exercised that the materials are perfectly clean and well washed. Only the best Portland cement, ground very fine, should be used, and the concrete must be well mixed. Traffic must be diverted, or the footway covered with planks, until the concrete is properly set. If the weather is hot and dry the concrete should be covered with sand and kept damp.

Concrete paths of this description cannot be laid in frosty weather and are difficult to repair when broken up. The necessity for the diversion of the traffic is also an objection. For these reasons concrete slabs or artificial stones are preferred, as they overcome the objections named.

A *monolith concrete footway* is laid in the following manner:—

Foundation.—The ground is excavated to a depth of about 6 in. below the finished level, and a bed of gravel 1 in. in thickness is then spread. A layer of clean hard stone (broken to pass all ways through a 3-in. ring) is next laid and well rolled, the surface being left about 2 in. below the finished level of the footway.

Concrete.—The footway is then divided into bays with strips of wood, and each alternate bay completed by laying in concrete, which is beaten or rolled into position.

The concrete should consist of one part Portland cement, two parts coarse *clean* gravel passed through a 1-in. screen, and two parts of clean sharp sand.

A finishing coat, 1 in. in thickness, of a finer and richer concrete is added before the former layer is set, and is well trowelled and smoothed to a finished surface. The finishing coat may consist of one part of cement to two parts granite chippings (to pass a $\frac{1}{4}$-in. sieve).

The battens are removed as the work is finished and the joints filled with fine sand. For concrete footways the transverse face should not exceed $\frac{3}{8}$ in. per foot.

Monolith concrete footways cost from 1s. 8d. up to 5s. 6d. per square yard, according to the locality and varying details in the mode of laying.

Imperial Stone Company in situ *Paving.*—The concrete is laid on a foundation of 4 in. of clinker or brick rubbish, in one coat 2 in. thick when finished; no surface coat is added, but the fine material is brought to the surface by trowel-tapping.

The footway is divided into bays from 4 ft. to 6 ft. in width by screeds formed on the foundation with wood strips (2 in. by $\frac{1}{4}$ in.), and each alternate bay is filled with granite concrete well worked and trowelled off smooth on the face. The intermediate bays are then filled in after the panels first formed have set, the wood strips being allowed to remain in so as to prevent the panels adhering. The surface of the paving is not indented, except upon inclines, as this is considered to interfere with its wearing qualities.

Stuart's Granolithic.—This paving as laid by Stuart's Granolithic Company consists of a bottom layer of Thames ballast concrete 2 in. in thickness laid upon a ¼-in. foundation of clinker, brick rubbish or broken stone.

A layer of granite concrete 1 in. in thickness is next laid, and well worked to remove cavities, and the surface trowelled to a fair face.

The surface is then sanded and rolled with a spiked roller, to indent the surface with a view of preventing slipping.

As in the case of the Imperial Stone Company above mentioned, the concrete is laid and finished off in panels, but the wood strips are omitted in the finished paving.

Guernsey Granite Concrete.—Guernsey granite concrete makes a good footway, and is laid 2 in. in thickness, upon a 4-in. foundation of gravel, furnace clinker, or other suitable material. Some 13,700 yards super. of this class of pavement were laid (1894) in Wimbledon. The following are the particulars of its construction*:—

"The concrete consisted of Guernsey granite chippings and Hilton & Anderson's cement; it was laid in two layers—the first, 1¼ in. thick, composed of chippings that passed through a ⅜-in. mesh and were retained on $\frac{3}{16}$-in. mesh, mixed with cement in the proportion of four of chippings to one of cement. For the top layer, ¾ in. thick, the chippings were finer, passing through a $\frac{3}{16}$-in. mesh, and cement was added in the proportion of one of cement to two of chippings. This layer was placed on the bottom layer within half an hour after the bottom was laid.

"Each layer of concrete was well worked after being placed in position, the surface of the top layer being trowelled off perfectly smooth. The channels were kept apart by flat iron bars, which were withdrawn after the concrete had been placed in position, and strips of wood 2 in. deep and ⅛ in. thick were then inserted before the face was trowelled off."

---

* "Proceedings of the Institution of Civil Engineers," vol. cxxii., part iv., p. 208.

BRICK PAVEMENTS.

Reference has already been made to the use of bricks for carriageways. In some towns they are also largely used for footway pavements, but cannot be recommended as an ideal pavement for general use. Only hard, vitrified bricks of the best quality should be used; ordinary bricks have been used where better material was not available at a reasonable price, but these are wholly unsuited for pavements.

Brick may advantageously be employed in back and narrow streets of towns in a brick-making locality, and when thoroughly vitrified form a very durable and cheap pavement. In West Bromwich a vitrified, Staffordshire brick pavement costs about 2s. 6d. per square yard, and has a life of over thirty years. In Southampton blue-brick pavement costs 3s. 6d. per yard super., including concrete foundation.

The objections to brick pavements are the multiplicity of joints, which make it difficult to cleanse; the variations in wear of adjacent bricks, thus forming depressions retaining puddles of water; also, bricks are rather difficult to bed to a level and uniform surface, and a bed of concrete is necessary for a satisfactory pavement. In lieu of this a foundation of 4 in. of clinker ashes has been used with success. Bricks, too, wear slippery and unevenly, and the skin is soon rubbed off unless of the best quality, thus exposing the interior of the brick, which wears rapidly. The appearance of a brick pavement is not in its favour and it is not a pleasant surface to walk upon.

Three kinds of bricks may be mentioned:—

Blue Staffordshire Bricks.—These are very largely used, and make a durable pavement when properly laid upon concrete, but the surface is so vitrified as to be slippery. At the same time the bricks are not burnt throughout, so that when the outer coating is worn through the red interior offers but little resistance to wear.

Buckley Bricks.—These are 10 in. by 5 in. by 2 in. deep, and form a good surface to walk over. They are more evenly burnt than Staffordshire bricks, the entire brick being usually

P

vitrified throughout; but the surface does not present the same vitrified, glossy surface, and retains sufficient roughness to afford a better foothold.

Imperishable Granite Vitrified Bricks are of a light buff colour, and are manufactured from granite clays found in Devon.

Carriageway Paving Bricks (9 in. by 4½ in. by 2⅜ in.) of good quality, adapted for the wear and tear of street traffic, are made by Messrs. Woolliscroft, of Hanley. Bricks of this class have been very largely employed for street paving in America with success.

TAR PAVEMENTS.

Tar pavements and tar macadam stand in great favour in suburban districts and pleasure towns, where the traffic is light, and the methods of manufacture and laying vary somewhat in detail, according to the locality. Reference has already been made to the subject under the head of " Tar Macadam."

The following is a method very generally adopted :—

Gravel or stone chippings are screened through sieves of 1¼-in., ¾-in., ½-in. and ¼-in. gauge, and afterwards heated on iron plates with fires underneath. When well dried and heated the following ingredients are mixed, boiled in iron cauldrons, and added while hot—*viz.*, 12 gallons of tar, ½ cwt. of pitch, 2 gallons of creosote, 1 ton of screened material (stone chippings).

This tar concrete is then spread in layers, putting the largest size gravel at the bottom, finishing with the smallest at the top, each layer being well rolled by means of an iron roller of about ½ ton weight.

The tarred material should be kept a month or two before use, so that they may be thoroughly soaked by the composition.

In storing this material it should be kept free from damp. It improves if kept for at least two years.

The chippings require to be thoroughly heated to insure *dryness*, so that the composition will adhere firmly.

If the material used is too hard it causes a "bumpy" path,

which is disagreeable to walk upon. Broken Kentish ragstone or limestone make the best pavements.

The best time to execute the work is spring or winter, provided it is dry; the heat of the sun draws the composition away from the stone on to the surface of the pavement.

No water must be allowed to get to the foundation or the tar concrete will be seriously affected.

On completion a little white spar, grit or stone dust is thrown on the surface and rolled in, which adds to the appearance, and the surface is sanded to prevent the tar adhering to the feet of pedestrians.

"Dressing"—*i.e.*, tarring and sanding the surface of the footways—is done during dry weather, the first summer after the laying of the pavement, and every three years afterwards. The tar used must be well seasoned, or refined tar, heated in a cauldron with a little pitch.

The surface of the footpath is swept, the hot tar applied, and a layer of dry, sharp sand about $\frac{1}{8}$ in. thick spread on the tar, so as to prevent it adhering to pedestrians' feet. By persons walking over the path the sand is forced into the tar, and forms a skin which preserves the life of the path.

The cost of the pavement is as follows:—

Scarborough, for *roadways*, exclusive of foundation, 2s. per yard super.; for *footways*, 1s. per yard super.

Banbury, tar pavement, 9d. per square yard.

Bath, tar pavement, 1s. 9d. per square yard, and the footways last in good repair for twenty years.

Hereford, tar pavement, 3s. per square yard.

Darlington, tar pavement, 2s. per square yard.

„ retopping tar pavements, 3d. per square yard.

Doncaster, tar pavement, 1s. 8d. per square yard; life, from twenty to thirty years.

Harwich, tar pavement, 1s. 4d. per square yard; life, from ten to twelve years.

Peterborough, tar pavement, 1s. 3d. per square yard; life, from fifteen to twenty years.

Windsor, tar pavement, 2s.; life, twenty years.

Wimbledon, tar pavement, 3 in. thick, 2s. 6d. per yard super.; annual maintenance ½d. per yard super.; life, ten years.

ASPHALTE FOR FOOTWAYS.

Compressed Asphalte.—This is used principally for carriageways, but where pedestrian traffic is great it may advantageously be employed for footpaths. As a carriageway pavement it has already been dealt with, and, as it is laid in a similar manner for footways, need not again be described, except to mention that the thickness in the latter case is only about 1 in. to 1½ in. The form of asphalte generally employed upon footways is that known as " mastic."

Mastic Asphalte.—This is a manufactured compound, consisting of natural asphalte, artificial bitumen and grit. Artificial bitumen is used because of the scarcity of natural bitumen. Trinidad pitch is its chief component, and to which is added about 6 per cent. of shale oil. This mixture is boiled for twenty-four hours, and the top liquid is ladled out, which is the artificial bitumen. Its quality may be tested by taking a piece between the fingers and drawing it out to a string; the quality is good if it does not snap until drawn out to a fine thread.

The mastic asphalte is prepared in the following manner: 5 per cent. to 7 per cent. of artificial bitumen, 20 per cent. to 30 per cent. of grit, and the balance in powdered asphalte, are placed in a large covered cauldron and heated or "cooked" for about five hours. The mixture liquefies at a temperature of from 280 deg. Fahr. to 300 deg. Fahr.

When it is to be used within, say, a 10-mile radius, it is run out into cauldrons on wheels (commonly called "locomobiles"), which are provided with a fire and with agitators worked by an endless chain attached to the axle of the wheels, and is thus transported direct to the site where it is to be spread. When used it should be hot enough to vaporise a drop of water. A test of its being ready and fit to lay is made by plunging a wooden spatula into it, which should come out without any of the asphalte adhering to it; also, it is observed to be ready when jets of light smoke dart out of the mixture. It is taken in pails, previously heated, and spread over the foundation by means of a float. Silver sand is spread over the surface and rubbed in by floats, and in about six hours the footway is ready for traffic.

It very frequently happens that mastic asphalte is to be laid at a distance from the works, in which case, instead of running it from the cauldrons into the "loco-mobiles," it is run into moulds and formed into flat cylindrical cakes of about 50 lb. weight each. These are taken to the site and are remelted in small round street cauldrons containing eight or twelve cakes each. The grit is added sometimes in the fixed cauldrons and sometimes in the street cauldrons. From 3 per cent. to 4 per cent. of additional bitumen is added to make up for the loss by evaporation. The process of laying is performed as above described.

One ton of asphalte covers about 20 square yards when laid 1 in. thick. A skilled workman can lay 140 square yards to 180 square yards in a day if properly assisted.*

The foundation for an asphalte footway should consist of 3 in. or 4 in. of Portland cement concrete (six to one). The surface of the concrete is smoothed over, and, after four days allowed for drying, the mastic asphalte is floated over its surface and the path completed.

It avoids movement, due to variations in temperature, &c., in the concrete foundation, resulting in unsightly cracks; the concrete should be sufficiently thick, and should be laid in sections, with the joints between them filled with some compressible substance. Compressed asphalte in footways is more liable to cracks than mastic, having no elasticity in itself, and when affected by the contracting force of the concrete is fractured, but when laid in streets of heavy traffic these cracks are not observable on the surface—the traffic welding the asphalte together again before they appear. In footways, also, of heavy traffic less cracks appear than in those of light traffic. It has been observed that these cracks are exactly of the shape and in the position of those in the concrete foundation underneath.

* "Proceedings of the Institution of Civil Engineers," vol. xliii., p. 293.

Compressed asphalte has a life about one-third longer than mastic asphalte under similar conditions.

The advantages of asphalte pavements are :—

(1) Durability.

(2) Impervious to moisture; cellars, &c., under footways are kept dry.

(3) Good foothold, and pleasant to walk upon; it also affords a smooth surface, unbroken by joints, and is easily cleansed.

(4) Is even and uniform in wear, and may be repaired with neatness.

Cost of Asphalte Footways.—This is to a very large extent a local question, and it can only be stated what the price has been in certain places.

Compressed or mastic asphalte in Chelsea,* 1 in. in thickness on 3 in. of concrete, cost (1887) 6s. 3d. per square yard. The same, but ¾ in. thick, was 5s. 6d. per square yard.

Compressed asphalte, ¾ in. thick, on ¼ in. of mastic asphalte, laid on 3 in. of concrete, cost 7s. per square yard.

Compressed or mastic asphalte, 1 in. thick, on existing concrete foundation (relay) cost 3s. 6d.

These prices carried a guarantee of free maintenance for ten years. The vestry prepared the foundation for the concrete in new work at a cost of 2d. per square yard.

The specification provided that the asphalte should be cut open at distances not exceeding 50 ft. apart and the thickness of the material measured. Five out of every six of such measurements were to be at least the specified thickness, and the average of every six to be at least the specified thickness.

In breaking a sample of asphalte pavement the affinity between the asphalte and the grit is found to be so great that the pieces of grit are found broken in half. The grit makes a durable footway and lessens the cost, but makes the asphalte more difficult to spread.

Inferior materials are sometimes substituted for natural

---

* "Proceedings of the Association of Municipal and County Engineers," vol. xiii.

asphalte, such as ground chalk, fireclay, and pitch or gas tar, or ground limestone mixed with bitumen.

CORK PAVEMENTS.

The Improved Cork Pavement Company has introduced patent cork bricks for paving purposes, and which are now in use for streets, roads, stables, yards, and such like. This class of pavement, it is claimed, is non-absorbent, non-slippery, practically noiseless, is far more durable than wood pavements, does not expand or contract when laid, and is a perfectly sanitary pavement.

The bricks for street paving are made 9 in. by $4\frac{1}{2}$ by 2 in., and cost, on rail at Barking station (near the company's works), 9s. 6d. per super. yard, including jointing material. The price for laying on buyer's prepared foundation in London is 1s. per square yard.

The pavement is said not to become slippery after rain.

For carriageways the bricks are laid upon a strong concrete foundation, 6 in. thick, rendered over, and formed to the intended contour of the street surface. The concrete is allowed to thoroughly set, and the surface then receives a thin coating of hot mineral tar, spread with a brush and then allowed to cool.

The jointing material, as sent with the bricks, is heated in an ordinary asphalte boiler, and care should be taken to keep it well stirred to prevent burning. It is then ladled into buckets as required, and only used so long as it is quite hot and thin. The bricks are dipped in the hot jointing material at such an angle that the entire bottom is covered and nearly the whole of the way up one side and one end of the brick. This is readily done by using a slater's hammer, or any similar tool, by sticking the point into the brick and then dipping the brick into the hot material, as described. The brick is then quicky and firmly placed in position, and the joints kept as thin as possible, but full of bitumen, and the surface of the brick at its proper level. Any bitumen dropped on the bricks during laying or any excess from the joints should be scraped off when hard. If the joints are properly filled with bitumen no grouting with sand and cement is required.

The Improved Cork Pavement Company has carried out the following road work, among others :—

London County Council: Belvedere-road (Lambeth), Newington Butts, Harrow-road.

Paddington Vestry: Chester-mews (Clarendon-place), Conduit-mews, &c.

Chelsea Vestry: Basil-street (Sloane-street, S.W.), laid 1897.

St. George's, Hanover-square, Vestry: Curzon-street, Mayfair, W. (1898).

Bournemouth Corporation: Cecil-road, Sea-road, Boscombe (January, 1898).

Nottingham Corporation: Lincoln-street.

MOVING PAVEMENTS.

This is one of the most recent novelties in pavements, and its latest development* is to be seen at St. Ouen, on the Seine, where it was introduced in February, 1899. Upon a platform, raised somewhat above the street level, two wooden tracks run side by side, one at the rate of 3 miles, the other at the rate of 5 miles, an hour. Passengers readily step on to the first and then pass on to the second without risk, but posts are placed at intervals to assist the timorous.

The paths themselves are built of planks of short length, so ingeniously jointed as to enable a sharp curve to be taken. These planks rest upon iron rails, to which motion is imparted by revolving wheels below. The system is, in fact, a sort of inversion of ordinary railway locomotion.

Moving pavements of this description are the means to be employed at the Paris Exhibition (1900) to transport visitors from point to point.

SELECTION OF PAVING MATERIAL.

In selecting the class of material to be adopted in paving the streets of any particular town it should be borne in mind that the question is materially affected by—

---

* *The Daily Chronicle.*

(*a*) The amount and description of the vehicular and pedestrian traffic.

(*b*) The inclination of the streets and the width of carriageway and footway.

(*c*) The nature of the buildings adjoining the street, whether business premises or residential property.

(*d*) The local facilities for obtaining a suitable material.

There is another point which must not be lost sight of—viz., the financial position of the authority for whom the proposed paving works are to be executed; but this, as a rule, may be left in the hands of the authority itself, the surveyor advising the adoption of the best material under the circumstances, keeping in view the points above mentioned.

It may be regarded as an ascertained fact that the pavement which is the most suitable for the traffic is the most economical in the end. It would be false economy to lay an inferior or second-rate pavement upon footways having a considerable amount of traffic and wear, as the maintenance would necessarily be great, and renewal would be required after a comparatively short period. At the same time a pavement which would be suitable for a heavy commercial traffic would be altogether out of place in a quiet provincial town or in a health resort.

The question for the selection of a paving material is materially affected by the gradient of the street. Stone setts may be used on streets not steeper than 1 in 16, wood on streets not steeper than 1 in 36, and asphalte would be unsuitable for streets steeper than 1 in 60.

Wood is unsuited to narrow streets, as the light and air should not be excluded from this class of pavement.

The continuity of pavements is also an important matter. Horses and drivers become accustomed and grow more confident in passing over continuous lengths of the same paving material, and accidents from the falling of horses are much more likely to occur where several different materials are employed in comparatively short lengths.

Granite sett pavements are suitable for streets containing many large warehouses; wood and asphalte pavements for streets lined with shop property, as in many parts of London; while tar-macadam, gravel or ordinary macadam may be very suitable for seaside health resorts or residential suburban districts with light traffic.

Local materials should not always be allowed to receive preference unless they are decidedly good.

The question as to whether a pavement is economical is affected by the life of the pavement, and upon which depends the first cost and interest, and the sinking fund on this cost. The cost of cleansing, reinstatement after disturbance, and the annual cost of maintenance, must all be taken into consideration. Land values and business interests are also, in a measure, affected thereby.

The effect of the introduction of a good carriageway pavement is readily to be observed in the increased loads which can be drawn by horses traversing the streets of a well-paved town, thus giving the public the benefit of a reduced cost in the transportation of goods.

The points to be sought after in selecting a good street pavement are—it should be durable, impervious to moisture, should afford a good foothold, be suited for laying upon all gradients, adapted to all classes of traffic, afford ease of traction, and be practically noiseless; it should be easily cleansed and repaired after removals, and should not create mud or dust, or be affected by the weather, heat, or other climatic changes. The pavement should also be economical in the cost of maintenance and in its first cost. The appearance of footway pavements also should receive consideration; this should not be too sombre, but be of an agreeable, uniform tint. For an ideal pavement, the fewer the joints the better.

Artificial tests for street-paving materials are not to be relied upon; the only reliable guide is an actual time test of the material laid in the public thoroughfares under the usual conditions of wear and tear.

THE CONSTRUCTION OF "NEW STREETS" UNDER THE BY-LAWS OF THE LOCAL GOVERNMENT BOARD.

The laying out and construction of new streets in the course of the development of districts under the government of local authorities is regulated by local by-laws prepared by those authorities under sec. 157 of the Public Health Act, 1875, which prescribes that every urban authority may make by-laws "with respect to the level, width and construction of new streets, and the provisions for the sewerage thereof."

The by-laws necessarily vary somewhat, according to local requirements, but, for the guidance of urban sanitary authorities, a series of model by-laws has been issued by the Local Government Board. These have been largely employed by local authorities throughout the country as models, each authority making certain modifications to meet the special needs of their particular district.

The usual stages in the development of a new street before it finally becomes a "highway repairable by the inhabitants at large" are as follows—the time having arrived for the development of a building estate, the projected new streets are laid out, constructed and sewered by the estate owners in conformity with the requirements of the local by-laws. The plots of land abutting upon such streets are very generally sold to builders and others, and in due course the street thus becomes built upon. Up to this point the street is known technically as a "private street," and, as the result of the cartage of quantities of heavy building materials over it, together with the wear and tear of the ordinary traffic using such street, coupled with the invariable neglect of its repair by the abutting owners, the street generally falls into a bad state of repair, and the occupiers of the houses, although called upon to pay their quotum of the general rates for purposes of lighting, street maintenance, &c., do not enjoy these privileges so far as their own particular road is concerned.

At this stage the local authority steps in and puts into force the powers conferred upon them by sec. 150 of the Public

Health Act, 1875, or, better, by the Private Street Works Act, 1892, enabling the authority to cause the street to be properly and effectually made up to their satisfaction at the cost of the abutting owners, and then to declare such street to be a " highway repairable by the inhabitants at large," from which time its maintenance is undertaken by the authority.

The *Model By-Laws* above mentioned and the methods of procedure under the Private Street Works Act, 1892, will now be dealt with in detail.

BY-LAWS OF THE LOCAL GOVERNMENT BOARD AS TO NEW STREETS.

(I.) Level of New Streets.—" Every person who shall lay out a new street shall lay out such street at such level as will afford the easiest practicable gradients throughout the entire length of such street for the purpose of securing easy and convenient means of communication with any other street or intended street with which such new street may be connected or may be intended to be connected, and as will allow of compliance with the provisions of any statute or by-law in force within the district for the regulation of new streets and buildings."

The Public Health Act, 1875, sec. 4, defines a street in the following manner :—

"'Street' includes any highway (not being a turnpike road), and any public bridge (not being a county bridge), and any road, lane, footway, square, court, alley or passage, whether a thoroughfare or not."

The term "street" applies to anything which is "a street in the ordinary sense of the term," although not a "highway," &c. As to the phrase "a street in the ordinary sense of the term," Brett, M.R., in *Mayor, &c., of Portsmouth* v. *Smith*, remarked "that he thought that the word 'street,' when popularly used, meant 'a thoroughfare bounded either on one side or both sides by houses.'"

It is not necessary that there should be houses on both sides of a thoroughfare in order to make a "street" (*Richards* v. *Kessick*).

Lord Selborne, in *Robinson* v. *Barton Local Board*, observed: "In the natural and popular sense of the word, 'street,' or the words 'new street,' I should certainly understand a roadway with buildings on each side (it is not necessary to say how far they must or may be continuous or discontinuous); and by 'new street' a place which before had not that character, but which, by the construction of buildings on each side, or possibly on one side, has acquired it."

"The Imperial Dictionary" gives the following definition of a "street," which was approved of by Jessel, M.R., in *Taylor* v. *Oldham Corporation*: "The street itself is no doubt properly the paved or prepared road, that is, the street. It sometimes includes the houses along each side of it. But that is not its proper meaning. It is called a street even without houses. There are some streets with no houses. But the usual common meaning of the word 'street' is a road with houses on one or both sides of it."

Further, in reference to the question of what is a new street, it was held, in *Baker* v. *Mayor, &c., of Portsmouth*, that the words "with respect to the level, width and construction of new streets" included the construction of the buildings and the buildings themselves, and front gardens or forecourts, or whatever else may be at the side of the roadway; while in *Maude* v. *Baildon Local Board* it was held to be a question of fact for the justices whether or not a road is a new street.

(II.) *Width and Construction of New Streets.*—"Every person who shall lay out a new street which shall be intended for use as a carriage road shall so lay out such street that the width thereof shall be (*thirty-six feet*) at least."

A width of 36 ft. may be regarded as the minimum width for a new street intended for carriage traffic in any district of an urban character or likely to develope to such. Such a street, with two footways 6 ft. in width each, provides a carriageway of 24 ft., which is sufficient to allow a vehicle to pass along the middle of the roadway when one vehicle is standing at each side of the way. A carriageway should be some multiple of 8 ft. in width, as vehicles can pass each other with ease in this space.

There is also a sanitary reason for securing, as far as possible, wide roadways, as it prevents the crowding of dwelling houses upon a given area, and so improves the character and public health of the district.

(*III.*) *Streets Exceeding* 100 *ft. in Length.*—" Every person who shall construct a new street which shall exceed (100 ft.) in length shall construct such street for use as a carriage road, and shall, as regards such street, comply with the requirements of every by-law relating to a new street intended for use as a carriage road."

The object of this section is to prevent the construction of narrow streets of indefinite length. It would, however, very probably be best to require all new streets to be of a minimum width, as prescribed in Clause II.

(*IV.*) *Width of New Street not intended for Carriage Traffic.* —" Every person who shall lay out a new street which shall be intended for use otherwise than as a carriage road, and shall not exceed in length (100 ft.), shall so lay out such street that the width therof shall be (24 ft.) at the least.

" Provided always that this by-law shall not apply in any case where a new street shall not be intended to form the principal approach or means of access to any building, but shall be intended for use solely as a separate means of access to any premises for the purpose of removing therefrom the contents of the receptacle of any privy, or of any ash-pit, or of any cesspool, without carrying such contents through any dwelling-house or public building, or any building in which any person may be, or may be intended to be, employed in any manufacture, trade or business."

New streets affording the principal approach to houses are allowed, under certain circumstances, by this clause of a width unsuited to the general vehicular traffic; but, it should be noted all *back streets* affording secondary access to houses are exempted. The provision of these cannot be enforced. Reference may be made to the case of *Waite* v. *Garston Local Board*, in which a by-law was held to be *ultra vires* which required that no dwelling-house should be erected without having at the rear or side a roadway not less than 12 ft. in width, communicating with an adjoining highway, and in such position as the local authority approved, for the purpose of affording efficient means of access to the **privy or ash-pit** belonging to such dwelling.

As to the *cleansing* of back streets, passages, &c., this may be required by the sanitary authority, and if not done to their satisfaction may be executed by them and charged to the respective occupiers (see Public Health Acts Amendment Act, 1890, sec. 27).

(*V.*) *Construction of New Streets.*—" Every person who shall construct a new street for use as a carriage road shall comply with the following requirements " :—

Width of Carriageway.—" He shall construct the carriageway of such street so that the width thereof shall be (24 ft.) at the least."

Surface of Carriageway.—" He shall construct the surface of the carriageway of such street so as to curve or face from the centre or crown of such carriageway to the channels at the sides thereof; the height of the crown of such carriageway above the level of the side channels being calculated at the rate of not less than ($\frac{3}{8}$ in.) and not more than ($\frac{3}{4}$ in.) for every foot of the width of such carriageway."

Width of Footways.—" He shall construct on each side of such street a footway of a width not less than (one-sixth) of the entire width of such street."

Surface of Footways.—" He shall construct each footway in such street so as to slope or fall towards the curb or outer edge at the rate of ($\frac{1}{2}$ in.) in every 1 ft. of width if the footway be not paved, flagged, or asphalted; and at the rate of not less than ($\frac{1}{4}$ in.) and not more than ($\frac{1}{2}$ in.) in every foot of width if the footway be paved, flagged, or asphalted."

Curbing.—" He shall construct each footway in such street so that the height of the curb or outer edge of such footway above the channel of the carriageway (except in the case of crossings, paved or otherwise, formed for the use of foot passengers) shall not be less than (3 in.) at the highest part of such channel and not more than (7 in.) at the lowest part of such channel."

The width of carriageway prescribed should be some multiple of 8 ft., so as to allow of a vehicle being easily driven past when one is standing on each side of the roadway.

The heights of the crown of the roadway, if the curb above the channel, as well as the inclinations of the footways and similar matters, must be regulated by local requirements and the classes of materials and pavements usually employed in the district.

In *Baker* v. *Portsmouth (Mayor, &c.)* a by-law which required that no building should be erected until the street had been constructed to the approval of the local authority was considered good by the Court of Appeal on the ground that "the construction of new streets" included the construction of the buildings by the sides of same.

As to the construction of a street, in the case of *Rudland* v. *Mayor, &c., of Sunderland,* a by-law requiring that "every person who constructs a new street shall cause the curb of each footpath in such street to be put in such level as may be fixed or approved by the urban sanitary authority," and that "no person shall commence the erection of a building in a new street unless and until the curb of each footpath therein shall have been put in pursuant to the precedent requirement," was held to be unreasonable, and consequently unenforceable, for the following reasons:—

"There was no limit of time or place; whether a new street was broad or narrow, short or extensive, the curb of each footpath must be laid on both sides before the owner of any portion of land on it could begin to build; the by-law was not even confined to the particular piece of land opposite to that on which the owner might be going to build; the by-law put it in the power of the urban sanitary authority to dictate to a man when he should begin to build; also, it was doubted whether the authority had power to make such provision with respect to matters specifically provided for—*viz.,* by sec. 150 (Public Health Act, 1875), since they amounted to imposing on landowners an absolute duty to do certain things under pain of incurring a penalty, while that section merely provides that if they decline to do them there shall be no penalty, but the authority shall themselves do the work and recover the expenses."*

(VI.) *Entrance to New Streets.*—" Every person who shall construct a new street shall provide at one end at least of such street an entrance of a width equal to the width of such street, and open from the ground upwards."

This clause prohibits the construction of streets with *culs de sac*, and applies to all new streets, whether intended for vehicular traffic or not. The entrance is required to be unimpeded by projections or obstructions of any kind.

In the case of *Hendon Local Board* v. *Pounce* this by-law was held to be reasonable and valid, and a " landowner was restrained by injunction from building on his land so as to form a ' bottle-neck' street, although he could not provide an entrance of the width required by the by-law on his own land."†

" The *entrance* to a new street was distinguished by Kekewich, J., from the ' mode of access' to the street in a later case where the entrance was from a narrow public road, and an interlocutory injunction was granted to restrain the landowner from constructing or commencing the new street until an entrance should have been provided according to the by-law (*Bromley Local Board* v. *Lloyd*); but at the trial, before Wills, J., the injunction was dissolved, the learned judge holding that the ' entrance' meant a practicable way into the street, and considering that the language in the report of the Hendon case did not sufficiently indicate what it was that North, J., did decide.†

The Metropolis Management Act, 1862, sec. 98, also requires the entrance of a " new street" to be " open from the ground upwards," similar to the above by-law, and in *Daw & Son* v. *London County Council* it was held that this provision prevented the erection of a barrier across the end of the street to exclude the public ; also, that a person could be convicted under this enactment as for a continuing offence.

---

* "The Law of Public Health," by W. S. Glen (Knight & Co.), eleventh edition, vol. i., p. 325.

† " The Law of Public Health," by W. S. Glen (Knight & Co.).

Sewerage of New Streets.—Sec. 157 of the Public Health Act, 1875, sub-sec. (1), in addition to by-laws "with respect to the level, width and construction of new streets," also authorises by-laws to be made containing "provisions for the sewerage thereof." The "Model By-Laws" of the Local Government Board, however, contain no such provisions, and the circumstances which led to their omission are explained in the following extract from a circular-letter of the Local Government Board dated July 25, 1877 :—

"It will be seen that the model series contains no by-laws specifying provisions for the sewerage of new streets; and the reason for this is that the conditions which such by-laws must satisfy are to so great an extent dependent upon the varying circumstances of different localities. The board do not anticipate that inconvenience will result from the absence of satisfactory by-laws with respect to sewerage, for it may be doubted whether any powers which under such by-laws may be lawfully assumed by sanitary authorities will, as regards extent and efficacy, compare with the powers which they derive from the express provisions of the Public Health Act."

MAKING UP OF STREETS UNDER THE PRIVATE STREET WORKS ACT, 1892.

Previous to the passing of the Private Street Works Act, 1892, the making up of "private streets" was carried out under secs. 150, 151 and 152 of the Public Health Act, 1875, sec. 152 of which (dealing with the "adoption of private streets") was subsequently repealed, and other provisions substituted in lieu thereof by sec. 41 of the Public Health Acts Amendment Act, 1890.

As the methods of procedure under the Private Street Works Act, 1892, are in very many respects preferable to those under the sections of the Public Health Act, 1875, above mentioned, this more recent enactment is that under which most of this class of work is now executed. The earlier Act therefore need not be referred to in detail, except to point out a few of the differences between it and the 1892 Act.

Under the 1875 Act the urban authority had to run the risk of abutting owners raising objections which may be fatal to their power of recovering the expenses of the carrying out of the works, or which may require a new apportionment after the cost of the work had been incurred and apportioned. Under sec. 7 of the 1892 Act an opportunity is afforded of taking the objections of owners before the expense of the works is incurred. If objections are not received, then the authority may proceed with the works without incurring any risk of such objections being subsequently made with success.

Under the Public Health Act, 1875, unless the amount in dispute is less than £20, any objections to the apportionment of the expenses are to be determined by arbitration; but under the 1892 Act such objections are dealt with by a court of jurisdiction without regard to the amount in dispute. Also, objections to the effect that the proposed works are insufficient or unreasonable, or that the expenses are excessive, are settled in the same manner instead of by the Local Government Board.

Sec. 9 (Act, 1892) provides for the inclusion of any incidental works necessary for bringing the street, as regards sewerage, drainage, level, or other matters, into conformity with any other streets, including the provision of separate sewers for the reception of sewage and of surface water respectively.

Secs. 19 and 20 provide for the adoption of private streets, and follow sec. 41 of the Act of 1890 in providing for the adoption of the maintenance of the street by the authority in cases where some and not all of the works have been carried out. The authority also may adopt a street and make it a highway without having regard to the wishes of the owners; and, further, the Act obliges them to adopt a street if a majority of the owners so require.

The last clause of sec. 150 of the Public Health Act, 1875, enables the authority to make up the whole street where part of it is already a highway. This is not included in the 1892

Q^2

Act, and one of the objections which may be taken under sec. 7 is "that a street, or part of a street, is (in whole or in part) a highway repairable by the inhabitants at large."

Under sec. 10 of the new Act the authority, in apportioning expenses, are not obliged to base such apportionments upon frontage alone, but they may, if they think fit, have regard to the greater or less degree of benefit derived by any premises from the proposed works, and to the amount and value of any work already done by any owners, and may also include premises which do not abut, &c., on the street, but access to which is obtained from the same through a court or passage.

By sec. 22 the expenses chargeable to abutting canal or railway companies under the 1875 Act, provided they have no direct communication with the street, are thrown upon the remaining abutting owners, unless and until a communication with the street is made from the premises of the company. The authority may, if they so determine, charge any part of the expenses under the Act upon the rates of the district.

Expenses for private street works under the Act of 1892 are recoverable by action of debt in the county court, although amounting to £50 or more; and the six months' limitation is apparently not applicable to an action in the county court under the 14th sec. of the Private Street Works Act, 1892, as it is under sec. 261 of the Public Health Act, 1875. Expenses are also recoverable in the High Court of Justice, in a court of summary jurisdiction, or as charges on the premises.

CHAPTER 57.

An Act to amend the Public Health Acts in relation to Private Street Improvement Expenses. A.D. 1892.

[28th June, 1892.]

BE it enacted by the Queen's most Excellent Majesty, by and with the advice and consent of the Lords Spiritual and Temporal, and Commons, in this present Parliament assembled, and by the authority of the same, as follows:

1. This Act may be cited as the Private Street Works Act, 1892, and shall be construed as one with the Public Health Acts, and shall extend only to England; and this Act and the Public Health Acts may be cited together as the Public Health Acts.
Short title, construction and extent.

2. This Act shall extend and apply to any urban sanitary district in which it is respectively adopted under the provisions of this Act.
Adoption of Act.

3. The following provisions shall have effect with regard to the adoption of this Act by urban authorities:
Adoption of Act by urban authorities.

    (1.) The adoption shall be by a resolution passed at a meeting of the urban authority; and one calendar month at least before such meeting special notice of the meeting, and of the intention to propose such resolution, shall be given to every member of the authority, and the notice shall be deemed to have been duly given to a member of it if it is either—

    (*a.*) Given in the mode in which notices to attend meetings of the authority are usually given; or

    (*b.*) Where there is no such mode, then signed by the clerk of the authority, and delivered to the member or left at his usual or last known place of abode in England, or forwarded by post in a prepaid registered letter, addressed to the member at his usual or last known place of abode in England.

(2.) Such resolution shall be published by advertisement in some one or more newspapers circulating within the district of the authority, and by causing notice thereof to be affixed to the principal doors of every church and chapel in the place to which notices are usually fixed, and otherwise in such manner as the authority think sufficient for giving notice thereof to all persons interested, and shall come into operation at such time not less than one month after the first publication of the advertisement of the resolution as the authority may by the resolution fix, and upon its coming into operation this Act shall extend to that district.

(3.) A copy of the resolution shall be sent to the Local Government Board.

(4.) A copy of the advertisement shall be conclusive evidence of the resolution having been passed, unless the contrary be shown; and no objection to the effect of the resolution on the ground that notice of the intention to propose the same was not duly given, or on the ground that the resolution was not sufficiently published, shall be made after three months from the date of the first publication of the advertisement.

Local Government Board may extend Act to rural districts.
4. The Local Government Board may declare that the provisions contained in this Act shall be in force in any rural sanitary district, or any part thereof, and may invest a rural sanitary authority with the powers, rights, duties, capacities, liabilities and obligations which an urban authority may acquire by adoption of this Act, in like manner and subject to the same provisions as they are enabled to invest rural sanitary authorities with the powers of urban sanitary authorities under the provisions of section two hundred and seventy-six of the Public Health Act, 1875.

Interpretation.
5. In this Act, if not inconsistent with the context,—

The expression "urban authority" means an urban sanitary authority under the Public Health Acts.

The expressions "urban sanitary district" and

"rural sanitary district" mean respectively an urban sanitary district and a rural sanitary district under the Public Health Acts, and "district" means the district of an urban sanitary authority or of a rural sanitary authority, as the case may require.

The expressions "surveyor," "lands," "premises," "owner," "drain," "sewer," have respectively the same meaning as in the Public Health Acts.

The expression "street" means (unless the context otherwise requires) a street as defined by the Public Health Acts, and not being a highway repairable by the inhabitants at large.

Words referring to "paving, metalling, and flagging" shall be construed as including macadamising, asphalting, gravelling, kerbing, and every method of making a carriageway or footway.

6.—(1.) Where any street or part of a street is not sewered, levelled, paved, metalled, flagged, channelled, made good, and lighted to the satisfaction of the urban authority, the urban authority may from time to time resolve with respect to such street or part of a street to do any one or more of the following works (in this Act called private street works); that is to say, to sewer, level, pave, metal, flag, channel, or make good, or to provide proper means for lighting such street or part of a street; and the expenses incurred by the urban authority in executing private street works shall be apportioned (subject as in this Act mentioned) on the premises fronting, adjoining, or abutting on such street or part of a street. Any such resolution may include several streets or parts of streets, or may be limited to any part or parts of a street. *Private street works*

(2.) The surveyor shall prepare, as respects each street or part of a street,—

(a.) A specification of the private street works referred to in the resolution, with plans and sections (if applicable);

(b.) An estimate of the probable expenses of the works;

(c.) A provisional apportionment of the estimated expenses among the premises liable to be charged therewith under this Act.

Such specification, plans, sections, estimate, and provisional apportionment shall comprise the particulars prescribed in Part I. of the Schedule to this Act, and shall be submitted to the urban authority, who may by resolution approve the same respectively with or without modification or addition as they think fit.

(3.) The resolution approving the specifications, plans, and sections (if any), estimates, and provisional apportionments, shall be published in the manner prescribed in Part II. of the Schedule to this Act, and copies thereof shall be served on the owners of the premises shown as liable to be charged in the provisional apportionment within seven days after the date of the first publication. During one month from the date of the first publication the approved specifications, plans and sections (if any), estimates, and provisional apportionments (or copies thereof certified by the surveyor), shall be kept deposited at the urban authority offices, and shall be open to inspection at all reasonable times.

Objections to proposed works.

7. During the said month any owner of any premises shown in a provisional apportionment as liable to be charged with any part of the expenses of executing the works may, by written notice served on the urban authority, object to the proposals of the urban authority on any of the following grounds; (that is to say,)

(*a.*) That an alleged street or part of a street is not or does not form part of a street within the meaning of this Act;

(*b.*) That a street or part of a street is (in whole or in part) a highway repairable by the inhabitants at large;

(*c.*) That there has been some material informality, defect, or error in or in respect of the resolution, notice, plans, sections, or estimate;

(*d.*) That the proposed works are insufficient or unreasonable, or that the estimated expenses are excessive;

(*e.*) That any premises ought to be excluded from or inserted in the provisional apportionment;

(*f.*) That the provisional apportionment is incorrect in respect of some matter of fact to be specified in the objection or (where the provisional apportionment is made with regard to other considerations than frontage as hereinafter provided) in respect of the degree of benefit to be derived by any persons, or the amount or value of any work already done by the owner or occupier of any premises.

For the purposes of this Act joint tenants or tenants in common may object through one of their number authorised in writing under the hands of the majority of such joint tenants or tenants in common.

8.—(1.) The urban authority at any time after the expiration of the said month may apply to a court of summary jurisdiction to appoint a time for determining the matter of all objections made as in this Act mentioned, and shall publish a notice of the time and place appointed, and copies of such notice shall be served upon the objectors; and at the time and place so appointed any such court may proceed to hear and determine the matter of all such objections in the same manner as nearly as may be, and with the same powers and subject to the same provisions with respect to stating a case, as if the urban authority were proceeding summarily against the objectors to enforce payment of a sum of money summarily recoverable, The court may quash in whole or in part or may amend the resolution, plans, sections, estimates. and provisional apportionments, or any of them, on the application either of any objector or of the urban authority. The court may also, if it thinks fit, adjourn the hearing amd direct any further notices to be given. *Hearing and determination of objections.*

(2.) No objection which could be made under this Act shall be otherwise made or allowed in any court proceeding or manner whatsoever.

(3.) The costs of any proceedings before a court of summary jurisdiction in relation to objections under this Act shall be in the discretion of the court, and the court shall have power, if it thinks fit, to direct that the whole or any part of such costs ordered to be paid by an objector or objectors shall be paid in the first instance by the urban authority, and charged as part of the expenses of the works on the premises of the objector or objectors in such proportions as may appear just.

Incidental works.

9.—(1.) The urban authority may include in any works to be done under this Act with respect to any street or part of a street any works which they think necessary for bringing the street or part of a street, as regards sewerage, drainage, level, or other matters, into conformity with any other streets (whether repairable or not by the inhabitants at large), including the provision of separate sewers for the reception of sewage and of surface water respectively.

(2.) The urban authority in any estimate of the expenses of private street works may include a commission not exceeding five pounds per centum (in addition to the estimated actual cost) in respect of surveys, superintendence, and notices, and such commission when received shall be carried to the credit of the district fund.

Apportionment of expenses.

10. In a provisional apportionment of expenses of private street works the apportionment of expenses against the premises fronting, adjoining, or abutting on the street or part of a street in respect of which the expenses are to be incurred shall, unless the urban authority otherwise resolve, be apportioned according to the frontage of the respective premises; but the urban authority may, if they think just, resolve that in settling the apportionment regard shall be had to the following considerations; (that is to say,)

(a.) The greater or less degree of benefit to be derived by any premises from such works;

(b.) The amount and value of any work already done by the owners or occupiers of any such premises.

They may also, if they think just, include any premises which do not front, adjoin, or abut on the street or part of a street, but access to which is obtained from the street through a court, passage, or otherwise, and which in their opinion will be benefited by the works, and may fix the sum or proportion to be charged against any such premises accordingly.

11. The urban authority may from time to time amend the specifications, plans, and sections (if any), estimates, and provisional apportionments for any private street works, but if the total amount of the estimate in respect of any street or part of a street is increased, such estimate and the provisional apportionment shall be published in the manner prescribed in Part II. of the Schedule to this Act, and shall be open to inspection at the urban authority offices at all reasonable times, and copies thereof shall be served on the owners of the premises affected thereby; and objections may be made to the increase and apportionment, and if made shall be dealt with and determined in like manner as objections to the original estimate and apportionment. *Amendment of plan, &c.*

12.—(1.) When any private street works have been completed, and the expenses thereof ascertained, the surveyor shall make a final apportionment by dividing the expenses in the same proportions in which the estimated expenses were divided in the original or amended provisional apportionment (as the case may be), and such final apportionment shall be conclusive for all purposes; and notice of such final apportionment shall be served upon the owners of the premises affected thereby; and the sums apportioned thereby shall be recoverable in manner provided by this Act, or in the same manner as private improvement expenses are recoverable under the Public Health Act, 1875, including the power to declare any such expenses to be payable by instalments. *Final apportionment and recovery of expenses.*

38 & 39 Vict., c. 55.

(2.) Within one month after such notice the owner of any premises charged with any expenses under such apportionment may, by a written notice to the urban authority, object to such final apportionment on the following grounds, or any of them :—

(*a*.) That the actual expenses have without sufficient reason exceeded the estimated expenses by more than fifteen per cent.

(*b*.) That the final apportionment has not been made in accordance with this section.

(*c*.) That there has been an unreasonable departure from the specification, plans, and sections.

(3.) Objections under this section shall be determinded in the same manner as objections to the provisional apportionment.

<small>Charge on premises.</small>
13.—(1.) Any premises included in the final apportionment, and all estates and interests from time to time therein, shall stand and remain charged (to the like extent and effect as under section two hundred and fifty-seven of the Public Health Act, 1875) with the sum finally apportioned on them, or if objection has been made against the final apportionment with the sum determined to be due as from the date of the final apportionment, with interest at the rate of four pounds per centum per annum, and the urban authority shall, for the recovery of such sum and interest, have all the same powers and remedies under the Conveyancing and Law of Property Act, 1881, and otherwise as if they were mortgagees having powers of sale and lease and of appointing a receiver.

(2.) The urban authority shall keep a register of charges under this Act and of the payments made in satisfaction thereof, and the register shall be open to inspection to all persons at all reasonable times on payment of not exceeding one shilling in respect of each name or property searched for, and the urban authority shall furnish copies of any part of such register to any person applying for the same on payment of such reasonable sum as may be fixed by the urban authority.

<small>Recovery of expenses summarily or by action.</small>
14. The urban authority, if they think fit, may from time to time (in addition and without prejudice to any other remedy), recover summarily in a court of summary jurisdiction, or as a simple contract debt by action in any court of competent jurisdiction, from the owner for the time being of any premises in respect of which any sum is due for expenses of private street works the whole or any portion of such sum, together with interest at a rate not exceeding four pounds per centum per annum, from the date of the final apportionment till payment thereof.

15. The urban authority, if they think fit, may at any time resolve to contribute the whole or a portion of the expenses of any private street works, and may pay the same out of the district fund or general expenses, district rate, or other rate out of which the general expenses incurred under the Public Health Act, 1875, are payable.

Contribution by urban authority to expenses.

16. The incumbent or minister or trustee of any church, chapel, or place appropriated to public religious worship, which is for the time being by law exempt from rates for the relief of the poor, shall not be liable to any expenses of private street works as the owner of such church, chapel, or place, or of any churchyard or burial ground attached thereto, nor shall any such expenses be deemed to be a charge on such church, chapel, or other place, or on such churchyard or burial ground, or to subject the same to distress, execution, or other legal process, but the proportion of expenses in respect of which an exemption is allowed under this section shall be borne and paid by the urban authority.

Exemption from expenses of incumbent of church.

17. All owners of buildings or lands, being persons who under the Lands Clauses Acts are empowered to sell and convey or release lands, may charge such buildings or lands with such sum as may be necessary to defray the whole or any part of any expenses which the owners of or any persons in respect of such buildings or lands for the time being are liable to pay under this Act and the expenses of making such charge, and for securing the repayment of such sum with interest may mortgage such buildings or lands to any person advancing such sum, but so that the principal due an any such mortgage shall be repaid by equal yearly or half-yearly payments within twenty years.

Power for limited owners to borrow for expenses.

18. The urban authority may from time to time, with the sanction of the Local Government Board, borrow, on the security of the district fund and general district rates or other rate out of which the general expenses incurred under the Public Health Act, 1875, are payable, moneys for the purpose of temporarily providing for expenses of private street works, and the powers of the urban authority to borrow under the Public Health

Power for urban authority to borrow for private street works

Acts shall be available as if the execution of private street works under this Act were one of the purposes of the Public Health Act, 1875.

<small>Adoption of private streets.</small>
19. Whenever all or any of the private street works in this Act mentioned have been executed in a street or part of a street, and the urban authority are of opinion that such street or part of a street ought to become a highway repairable by the inhabitants at large, they may by notice to be fixed up in such street or part of a street declare the whole of such street or part of a street to be a highway repairable by the inhabitants at large, and thereupon such street or part of a street as defined in the notice shall become a highway repairable by the inhabitants at large.

<small>On street being paved, &c., urban authority to declare same public highway.</small>
20. If any street is now or shall hereafter be sewered, levelled, paved, metalled, flagged, channelled, and made good (all such works being done to the satisfaction of the urban authority), then, on the application in writing of the greater part in value of the owners of the houses and land in such street, the urban authority shall, within three months from the time of such application, by notice put up in such street declare the same to be a highway repairable by the inhabitants at large, and thereupon such street shall become a highway repairable by the inhabitants at large.

<small>Separate accounts of expenses of works.</small>
21.—(1.) The urban authority shall keep separate accounts of all moneys expended and recovered by them in the execution of the provisions of this Act in respect of street works, shall be applied in repayment of moneys borrowed for the purpose of executing private street works, or if there is no such loan outstanding then in such manner as may be directed by the Local Government Board.

<small>Railways and canals abutting but not communicating with streets not to be chargeable with private street expenses.</small>
22. No railway or canal company shall be deemed to be an owner or occupier for the purposes of this Act in respect of any land of such company upon which any street shall wholly or partially front or abut, and which shall at the time of the laying out of such street be used by such company solely as a part of their line of railway, canal, or siding, station, towing path, or works, and shall have no direct communication with such street; and the expenses incurred by the urban authority under the powers

of this Act which, but for this provision, such company would be liable to pay, shall be repaid to the urban authority by the owners of the premises included in the apportionments, and in such proportion as shall be settled by the surveyor; and in the event of such company subsequently making a communication with such street they shall, notwithstanding such repayment as last aforesaid, pay to the urban authority the expenses which, but for the foregoing provision, such company would in the first instance have been liable to pay, and the urban authority shall divide among the owners for the time being included in the apportionment the amount so paid by such company to the urban authority, less the costs and expenses attendant upon such division, in such proportion as shall be settled by the surveyor, whose decision shall be final and conclusive. This section shall not apply to any street existing at the date of the adoption of this Act.

23. All expenses incurred or payable by an urban authority and a rural sanitary respectively in the execution of this Act, and not otherwise provided for, may be charged and defrayed as part of the expenses incurred by them respectively in the execution of the Public Health Acts. *Expenses of local authority.*

24. All powers given to a local authority under this Act shall be deemed to be in addition to and not in derogation of any other powers conferred upon such local authority by any Act of Parliament, law, or custom, and such other powers may be exercised in the same manner as if this Act had not been passed. *Powers of Act cumulative.*

25. Neither sections one hundred and fifty, one hundred and fifty-one, and one hundred and fifty-two of the Public Health Act, 1875, nor section forty-one of the Public Health Acts Amendment Act, 1890, shall apply to any district or part of a district in which this Act is in force. *Certain sections of Public Health Acts not to apply.*

26. This Act shall not extend to prejudice or derogate from the estates, rights, and privileges of the Conservators of the River Thames, or render them liable to any charges or payments in respect of any of their works on or upon the shores of the River Thames. *For protection of Conservators of the River Thames.*

THE SCHEDULE.

PRIVATE STREET WORKS.

PART I.

PARTICULARS TO BE STATED IN SPECIFICATIONS, PLANS AND SECTIONS, ESTIMATES, AND PROVISIONAL APPORTIONMENTS.

Specifications.—These shall describe generally the works and things to be done, and in the case of structual works shall specify as far as may be the foundation, form, material, and dimensions thereof.

Plans and Sections.—These shall show the constructive character of the works, and the connections (if any) with existing streets, sewers, or other works, and the lines and levels of the works, subject to such limits of deviation (if any) as shall be indicated on the plans and sections respectively.

Estimates.—These shall show the particulars of the probable cost of the whole works, including the commission provided for by this Act.

Provisional Apportionments.—These shall state the amounts charged on the respective premises and the names of the respective owners, or reputed owners, and shall also state whether the apportionment is made according to the frontage of the respective premises or not, and the measurements of the frontages, and the other considerations (if any) on which the apportionment is based.

PART II.

PUBLICATION OF NOTICE.

Any resolution, notice, or other document required by this Act to be published in the manner prescribed by this schedule shall be published once in each of two successive weeks in some local newspaper circulating within the district, and shall be publicly posted in or near the street to which it relates once at least in each of three successive weeks.

INDEX.

	PAGE
Asphalte Carriageways, Advantages of	199
,, Introduction of	22
,, Pavements	...196, 227
,, Rock	196
Barrelling of Country Roads	11
Bethell's Process of Treating Timber	173
Blythe's ,, ,, ,,	176
Bog Roads	37
Boulton's Process of Treating Timber	174
Boncherie's ,, ,, ,,	178
Burnett's ,, ,, ,,	178
By-Laws as to Footways	205
,, ,, New Streets	235
Channelling of Carriageways	207
Concrete *in situ*	222
Cork Pavements	230
Cost of Asphalte Footways	230
,, Construction of Macadamised Roads	128
,, Tar Pavements	227
Creosoting Timber	170
Curbing	207
Drainage of Roads	51
Earthwork, Calculation of	49
,, Cutting and Embanking	38
,, Natural Slope	42
,, Stability	42
Embanking through Flooded Grounds	35
,, ,, ,, Soft ,,	37
,, Concavely and Convexly	40

R

	PAGE
Ermine-street	4
Fencing of Highways	64
Fireclay Curbs	210
Flagging, Adamant Stone	220
,, Caithness ,,	217
,, Imperial ,,	221
,, Lazonby ,,	218
,, Limestone	218
,, Pennant ,,	217
,, Purbeck ,,	217
,, Victoria ,,	218
,, York ,,	214
Flints	102
Footways	203
,, Materials for paving	213
Fosse Way	4
French Roads	4, 122
Gardner's Process for Preserving Wood	1
Geology	29
Gradient Ruling	33, 87
Granite	90
Granolithic, Stuart's	224
Gravel	102
Grout Cement	190
,, Pitch	190
Gullies for Streets	62
,, Crosta's	63
,, Sykes'	63
,, Victoria Stone Company's	64
,, Wakefield's	63
Hedges, Quickset	65
Highgate, Archway-road	118
Ikenild Roadway	4
Improved Wood Pavement	22
Iron Curbs	210
Joints in Wood Pavements	186
Kyan's Process of Treating Wood	179
Limestone	97

		PAGE
Macadam, J. L., ...		5, 12
Macneil, Sir J.		14
Margary's Process of Treating Wood		179
Materials for Roads		88
Metcalf, J....		11
Moving Pavements		232
Pavements, Asphaltic Wood		179
,,	Brick	...153, 225
,,	Carey's Wood	180
,,	Croskey's ,, ...	180
,,	Ellis's ,, ...	180
,,	Gabriel's ,, ...	181
,,	Harrison's ,, ...	181
,,	Henson's ,, ...	181
,,	Improved ,, ...	181
,,	in Liverpool...	145
,,	Joints in Wood	186
,,	Ligno-Mineral	182
,,	Lloyd's Wood	182
,,	Marshall's Wood	182
,,	Mowlem & Co.'s Wood	183
,,	Moving	232
,,	McDougall's	160
,,	Nicholson's Wood	183
,,	Norton's ,, ...	183
,,	Prosser's ,, ...	183
,,	Shiel's ,, ...	184
,,	Stone	142
,,	Stone's Wood	184
,,	Stowe's ,, ...	184
,,	Tar	226
,,	Wilson's Wood	184
,,	Wear of	194
,,	Wood	161
Payne's Process of Treating Wood		179
Private Streets, Making up of		...235, 242
,, ,, Works Act, 1892		245
Quenast Granite		129

	PAGE
Repose, Angle of	42
Retaining Walls	46
Rolling Roads	17, 131
Roman Roads	2
Sandstone	100
Section of Roads	82
Sidelong Ground	44
Simpson's Rule	49
Slabs, Granite	214
Specification, Compressed Asphalte	202
,, Granite Pavement (Liverpool)	145
,, Wood Pavements	192
,, York Paving	216
Stone Pavements	19
Syenite	95
Tar Macadam	125
Telford, Thomas	1, 13, 32
Timber, Creosoting	170
Traction	66, 79
Trap Rocks	95
Tredgold	1, 25
Town Roads	140
Watling-street	3
Wear of Wood Pavements	194
Width of Roads	79
Wind Velocity	67
Woods Employed in Street Paving	164
Wood Pavements	21
Yorkshire Flagging	214

www.ingramcontent.com/pod-product-compliance
Lightning Source LLC
Chambersburg PA
CBHW031935230426
43672CB00010B/1930